I0044423

Soil Science

Soil Science

Demi Wright

R CALLISTO REFERENCE

www.callistoreference.com

Callisto Reference,
118-35 Queens Blvd., Suite 400,
Forest Hills, NY 11375, USA

Visit us on the World Wide Web at:
www.callistoreference.com

© Callisto Reference, 2022

This book contains information obtained from authentic and highly regarded sources. All chapters are published with permission under the Creative Commons Attribution Share Alike License or equivalent. A wide variety of references are listed. Permissions and sources are indicated; for detailed attributions, please refer to the permissions page. Reasonable efforts have been made to publish reliable data and information, but the authors, editors and publisher cannot assume any responsibility for the validity of all materials or the consequences of their use.

ISBN: 978-1-64116-537-2 (Hardback)

Trademark Notice: Registered trademark of products or corporate names are used only for explanation and identification without intent to infringe.

Cataloging-in-Publication Data

Soil science / Demi Wright.
   p. cm.
Includes bibliographical references and index.
ISBN 978-1-64116-537-2
1. Soil science. 2. Soils. 3. Crops and soils. 4. Agriculture. I. Wright, Demi.
S591 .S65 2022
631.4--dc23

TABLE OF CONTENTS

It is with great pleasure that I present this book. It has been carefully written after numerous discussions with my peers and other practitioners of the field. I would like to take this opportunity to thank my family and friends who have been extremely supporting at every step in my life.

Soil science includes the study of soil as a natural resource that is found on the Earth's surface. It primarily focuses on the classification of soil and soil formation. Soil science also delves into the biological, physical and chemical properties of soil as well as its fertility. The classification of soil is based on the soil morphology. There are various branches of soil science such as pedology and edaphology. Pedology deals with the formation, morphology, chemistry and classification of soil. Edaphology focuses on the interaction of soil with living things. It is an umbrella field which encompasses and utilizes various disciplines including ecology, agronomy, geology, archaeology, silviculture, microbiology, physical geography and biology. This book explores all the important aspects of soil science in the present day scenario. Some of the diverse topics covered herein address the varied branches that fall under this category. Coherent flow of topics, student-friendly language and extensive use of examples make this book an invaluable source of knowledge.

The chapters below are organized to facilitate a comprehensive understanding of the subject:

Chapter – Introduction

The mixture of organic matter, minerals, liquids, gases and organisms that together support life is referred to as soil. The study of soil as a natural resource on the Earth's surface is known as soil science. It includes soil formation, classification and mapping. This is an introductory chapter which will briefly introduce all these significant aspects of soil science.

Chapter – Types of Soil

There are various types of soils which have different origins, compositions, functions and different uses. These can be categorized into sandy soil, silt soil, clay soil and loam soil. The topics elaborated in this chapter will help in gaining a better perspective about these various types of soil.

Chapter – Fundamental Concepts of Soil Science

Some of the most important fundamental concepts of soil science are soil fertility, soil respiration, soil functions, soil water retention, soil genesis, soil morphology, soil ecology, soil biology, humus and soil horizon. This chapter has been carefully written to provide an easy understanding of these fundamental concepts of soil science.

Chapter – Branches of Soil Science

There are various branches of soil science which deals with the constitution, properties and processes taking place in soil. Some of them are soil physics, soil chemistry, soil biology, soil micro-biology, soil ecology, pedology, edaphology, etc. All these diverse branches of soil science have been carefully analyzed in this chapter.

Chapter – Physical Properties of Soil

Some of the aspects that come under the physical properties of soil are soil texture, soil color, soil structure, soil porosity, soil density, soil texture and soil resistivity. This chapter has been carefully written to provide an easy understanding of these various physical properties of soil.

Chapter – Soil Quality

The ability of a soil to perform its functions which are necessary for environment and people is referred to as soil quality. Some of the aspects which are studied under the domain of soil quality include soil pH, soil acidification, soil organic matter and soil salinity. This chapter closely examines these aspects of soil quality to provide an extensive understanding of the subject.

Chapter – Soil Management

The application of practices, operations, and treatments for the protection of soil and enhancement of its performance, such as soil fertility or soil mechanics, is known as soil management. Some of the commonly used practices and techniques of soil management are crop rotation, nutrient management, tillage, green manuring, cover crop etc. The different practices and techniques of soil management have been thoroughly discussed in this chapter.

Chapter – Soil Pollution and Conservation

The presence of high concentrations of toxic chemicals in soil that can pose risk to human health and the ecosystem is known as soil pollution. The prevention of soil loss from erosion and reduced soil fertility caused by acidification is known as soil conservation. The topics elaborated in this chapter will help in gaining a better perspective about soil pollution and soil conservation.

Demi Wright

Introduction

The mixture of organic matter, minerals, liquids, gases and organisms that together support life is referred to as soil. The study of soil as a natural resource on the Earth's surface is known as soil science. It includes soil formation, classification and mapping. This is an introductory chapter which will briefly introduce all these significant aspects of soil science.

SOIL

Soil is the biologically active, porous medium that has developed in the uppermost layer of Earth's crust. Soil is one of the principal substrata of life on Earth, serving as a reservoir of water and nutrients, as a medium for the filtration and breakdown of injurious wastes and as a participant in the cycling of carbon and other elements through the global ecosystem. It has evolved through weathering processes driven by biological, climatic, geologic and topographic influences.

Since the rise of agriculture and forestry in the 8th millennium BCE, there has also arisen by necessity a practical awareness of soils and their management. In the 18th and 19th centuries the Industrial Revolution brought increasing pressure on soil to produce raw materials demanded by commerce, while the development of quantitative science offered new opportunities for improved soil management. The study of soil as a separate scientific discipline began about the same time with systematic investigations of substances that enhance plant growth. This initial inquiry has expanded to an understanding of soils as complex, dynamic, biogeochemical systems that are vital to the life cycles of terrestrial vegetation and soil-inhabiting organisms and by extension to the human race as well.

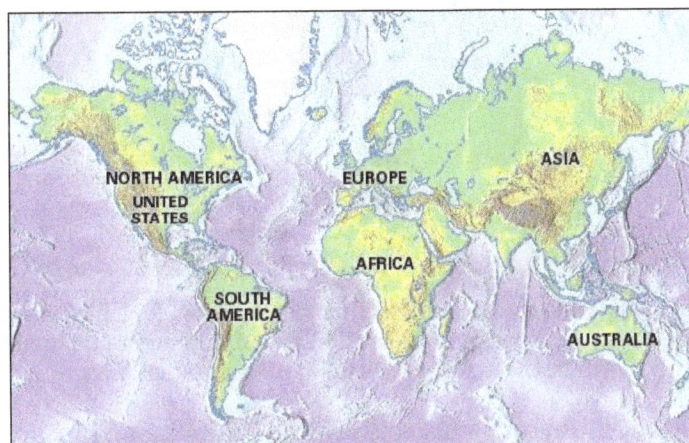

Interactive soil map of the world.

The Soil Profile

Soil Horizons

Soils differ widely in their properties because of geologic and climatic variation over distance and time. Even a simple property, such as the soil thickness, can range from a few centimetres to many metres, depending on the intensity and duration of weathering, episodes of soil deposition and erosion and the patterns of landscape evolution. Nevertheless, in spite of this variability, soils have a unique structural characteristic that distinguishes them from mere earth materials and serves as a basis for their classification: a vertical sequence of layers produced by the combined actions of percolating waters and living organisms.

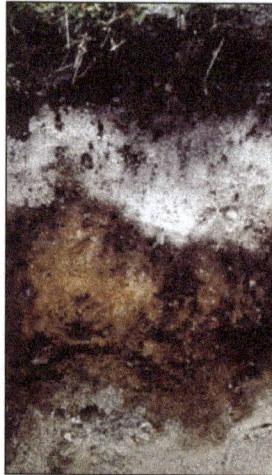

Podzol soil profile from Ireland, showing a bleached layer from which humus and metal oxides have been leached and subsequently deposited in the typically reddish horizon.

These layers are called horizons and the full vertical sequence of horizons constitutes the soil profile. Soil horizons are defined by features that reflect soil-forming processes. For instance, the uppermost soil layer (not including surface litter) is termed the A horizon. This is a weathered layer that contains an accumulation of humus (decomposed, dark-coloured, carbon-rich matter) and microbial biomass that is mixed with small-grained minerals to form aggregate structures.

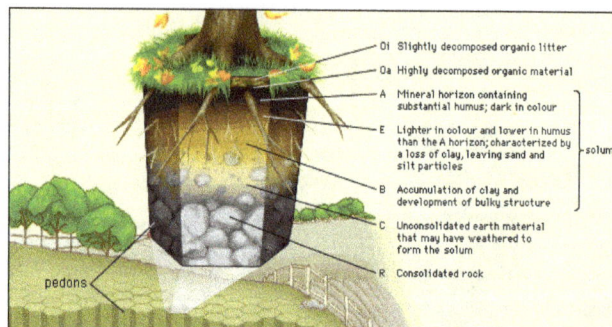

The soil profile, showing the major layers from the O horizon (organic material) to the R horizon (consolidated rock). A pedon is the smallest unit of land surface that can be used to study the characteristic soil profile of a landscape.

Below A lies the B horizon. In mature soils this layer is characterized by an accumulation of clay (small particles less than 0.002 mm [0.00008 inch] in diameter) that has either been deposited out of percolating waters or precipitated by chemical processes involving dissolved products of weathering. Clay endows B horizons with an array of diverse structural features (blocks, columns, and prisms) formed from small clay particles that can be linked together in various configurations as the horizon evolves.

Below the A and B horizons is the C horizon, a zone of little or no humus accumulation or soil structure development. The C horizon often is composed of unconsolidated parent material from which the A and B horizons have formed. It lacks the characteristic features of the A and B horizons and may be either relatively unweathered or deeply weathered. At some depth below the A, B and C horizons lies consolidated rock, which makes up the R horizon.

These simple letter designations are supplemented in two ways. First, two additional horizons are defined. Litter and decomposed organic matter (for example, plant and animal remains) that typically lie exposed on the land surface above the A horizon are given the designation O horizon, whereas the layer immediately below an A horizon that has been extensively leached (that is, slowly washed of certain contents by the action of percolating water) is given the separate designation E horizon, or zone of eluviation. The development of E horizons is favoured by high rainfall and sandy parent material, two factors that help to ensure extensive water percolation. The solid particles lost through leaching are deposited in the B horizon, which then can be regarded as a zone of illuviation.

The combined A, E, B horizon sequence is called the solum. The solum is the true seat of soil-forming processes and is the principal habitat for soil organisms. (Transitional layers, having intermediate properties, are designated with the two letters of the adjacent horizons.)

The second enhancement to soil horizon nomenclature is the use of lowercase suffixes to designate special features that are important to soil development. The most common of these suffixes are applied to B horizons: g to denote mottling caused by waterlogging, h to denote the illuvial accumulation of humus, k to denote carbonate mineral precipitates, o to denote residual metal oxides, s to denote the illuvial accumulation of metal oxides and humus, and t to denote the accumulation of clay.

Pedons and Polypedons

Soils are natural elements of weathered landscapes whose properties may vary spatially. For scientific study, however, it is useful to think of soils as unions of modules known as pedons. A pedon is the smallest element of landscape that can be called soil. Its depth limit is the somewhat arbitrary boundary between soil and "not soil" (e.g., bedrock). Its lateral dimensions must be large enough to permit a study of any horizons present—in general, an area from 1 to 10 square metres (10 to 100 square feet), taking into account that a horizon may be variable in thickness or even discontinuous. Wherever horizons are cyclic and recur at intervals of 2 to 7 metres (7 to 23 feet), the pedon includes one-half the cycle. Thus, each pedon includes the range of horizon variability that occurs within small areas. Wherever the cycle is less than 2 metres, or wherever all horizons are continuous and of uniform thickness, the pedon has an area of 1 square metre.

Soils are encountered on the landscape as groups of similar pedons, called polypedons, that contain sufficient area to qualify as a taxonomic unit. Polypedons are bounded from below by "not soil" and laterally by pedons of dissimilar characteristics.

Soil Behaviour: Physical Characteristics

Grain size and Porosity

The grain size of soil particles and the aggregate structures they form affect the ability of a soil to transport and retain water, air, and nutrients. Grain size is classified as clay if the particle diameter is less than 0.002 mm (0.0008 inch), as silt if it is between 0.002 mm (0.0008 inch) and 0.05 mm (0.002 inch) or as sand if it is between 0.05 mm (0.002 inch) and 2 mm (0.08 inch). Soil texture refers to the relative proportions of sand, silt and clay particle sizes, irrespective of chemical or mineralogical composition. Sandy soils are called coarse-textured, and clay-rich soils are called fine-textured. Loam is a textural class representing about one-fifth clay, with sand and silt sharing the remainder equally.

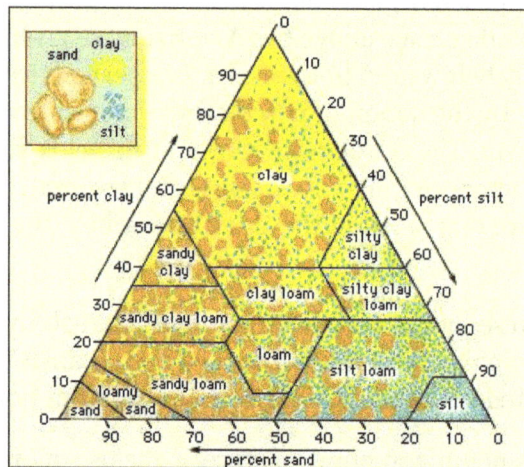

Soil texture as a function of the proportion of sand, silt, and clay particle sizes.

Pore radii (space between soil particles) can range from millimetre-scale between sand grains to micrometre-scale between clay grains. Soil particles falling into the three principal size categories may have various mineralogical or chemical compositions, although sand particles often are composed of quartz and feldspars, silt particles often are micaceous, and clay particles often contain layer-type aluminosilicates (the so-called clay minerals). Organic matter and amorphous mineral matter also are important constituents of soil clay particles.

Porosity reflects the capacity of soil to hold air and water and permeability describes the ease of transport of fluids and their dissolved components. The porosity of a soil horizon increases as its texture becomes finer, whereas the permeability decreases as the average pore size becomes smaller. Small pores not only restrict the passage of matter, but they also bring it into close proximity with chemical binding sites on the particle surface that can slow its movement. Clay and humus affect both soil porosity and permeability by binding soil grains together into aggregates, thereby creating a network of larger pores (macropores) that facilitate the movement of water. Plant roots open pores between soil aggregates, and cycles of wetting and drying create channels that allow water to pass easily. (However, this structure collapses under waterlogging conditions.) The stability of aggregates increases with humus content, especially humus that originates from grass vegetation. For soils that are not disturbed significantly by human activities, however, the pore space and the varieties of macropores are more important determinants of porosity than the soil texture. As a general rule, average pore size decreases from certain agricultural practices and other human uses of soil.

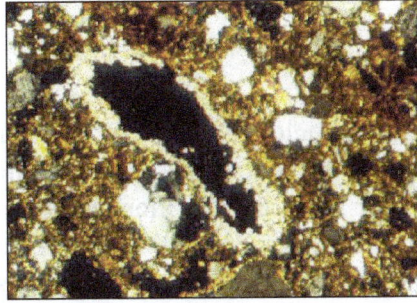

Microscopic view of an Inceptisol, showing small crystallites of carbonate minerals (around the central black void), quartz sand grains (white), and iron oxides and organic matter (dark brown).

Water Runoff

Aggregates of soil particles whose formation has not been influenced by human intervention are called peds. The peds in the surface horizons of soils develop into clods under the effects of cultivation and the traffic of urbanization. Soils whose a horizon is dense and unstructured increase the fraction of precipitation that will become surface runoff and have a high potential for erosion and flooding. These soils include not only those whose peds have been degraded but also coarse-textured soils with low porosity, particularly those of arid regions.

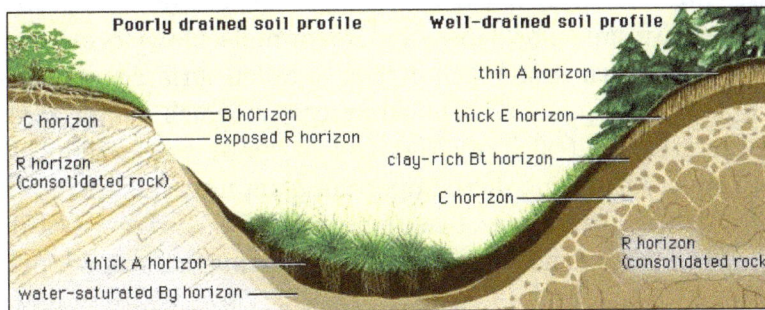

Soil profiles on hillslopes the thickness and composition of soil horizons vary with position on a hillslope and with water drainage.

For example, on the upper slopes of poorly drained profiles, underlying rock may be exposed by surface erosion, and nutrient-rich soils (A horizon) may accumulate at the toe slope. On the other hand, in well-drained profiles under forest cover, the leached layers (E horizon) may be relatively thick and surface erosion minimal.

A well-developed clay horizon (Bt) presents a deep-lying obstacle to the downward percolation of water. Subsurface runoff cannot easily penetrate the clay layer and flows laterally along the horizon as it moves toward the stream system. This type of runoff is slower than its erosive counterpart over the land surface and leads to water saturation of the upper part of the soil profile and the possibility of gravity-induced mass movement on hillslopes (e.g., landslides). It is also responsible for the translocation (migration) of dissolved products of chemical weathering down a hillslope sequence of related soil profiles (a toposequence). Subsurface water flow is also influenced by macropores, which, as noted above, are created through plant root growth and decay, animal burrowing activities, soil shrinkage while drying, or fracturing. In general, subsurface runoff processes are characteristic of soils in humid regions, whereas surface runoff is characteristic of arid regions and, of course, any landscape altered significantly by cultivation or urbanization.

Chemical Characteristics

Mineral Content

The bulk of soil consists of mineral particles that are composed of arrays of silicate ions (SiO_4^{4-}) combined with various positively charged metal ions. It is the number and type of the metal ions present that determine the particular mineral. The most common mineral found in Earth's crust is feldspar, an aluminosilicate that contains sodium, potassium, or calcium (sometimes called bases) in addition to aluminum ions. Weathering breaks up crystals of feldspars and other silicate minerals and releases chemical compounds such as bases, silica, and oxides of iron and aluminum (Fe_2O_3 and alumina [Al_2O_3]). After the bases are removed by leaching, the remaining silica and alumina combine to form crystalline clays.

The kind of crystalline clay produced depends on leaching intensity. Prolonged leaching leaves little silica to combine with alumina and results in what are known as 1:1 clays, consisting of alternating silica and alumina sheets; less extensive leaching leads to the formation of 2:1 clays, consisting of one alumina sheet sandwiched between two silica sheets. In neither case is the result solely one of the two types, though 1:1 clay is predominant in the tropics after prolonged leaching and 2:1 clay more abundant when leaching is less extensive in more temperate climates.

The solid soil particles are chemically reactive because of the presence of electrically charged sites on their surfaces. If a reactive site binds a dissolved ion or molecule to form a stable unit, a "surface complex" is said to exist. The formation reaction itself is called surface complexation. Surface complexation is an example of adsorption, a chemical process in which matter accumulates on a solid particle surface. Ions such as Ca^{2+} (calcium), Mg^{2+} (magnesium), Na^+ (sodium), and NO_3^- (nitrate) do not tend to adsorb strongly, making these important plant nutrients susceptible to easy replacement. Once ejected from their surface sites, these ions may be leached downward by percolating water to become removed from the biogeochemical cycles occurring in the upper part of the soil profile.

Freshwater leaching of soils brings hydrogen ions (H^+) that increase mineral solubility, releasing Al^{3+} (aluminum), a toxic ion that can displace nutrients such as Ca^{2+}. The gradual loss of nutrients and the accumulation of adsorbed H^+ and Al^{3+} characterize the build up of soil acidity, with its harmful effects on organisms. Soils display their acidity by a decrease in content of acid-soluble minerals (for example, feldspars or clay minerals) and an increase in insoluble minerals (iron and aluminum oxides). Soils weathered by freshwater leaching evolve from clay particles with a prevalence of metal ion-binding sites to highly weathered metal oxides that do not have sites that bind readily with metal ions.

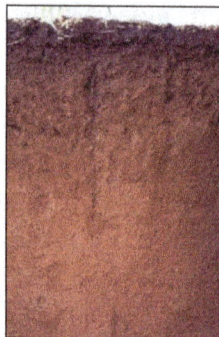

Ferralsol soil profile from Brazil, showing a deep red subsurface horizon resulting from accumulations of iron and aluminum oxides.

Organic Content

The second major component of soils is organic matter produced by organisms. The total organic matter in soil, except for materials identifiable as undecomposed or partially decomposed biomass, is called humus. This solid, dark-coloured component of soil plays a significant role in the control of soil acidity, in the cycling of nutrients, and in the detoxification of hazardous compounds. Humus consists of biological molecules such as proteins and carbohydrates as well as the humic substances (polymeric compounds produced through microbial action that differ from metabolically active compounds).

Mollisol soil profile, showing a typically dark surface horizon rich in humus.

The processes by which humus forms are not fully understood, but there is agreement that four stages of development occur in the transformation of soil biomass to humus:

- Decomposition of biomass into simple organic compounds.

- Metabolization of the simple compounds by microbes.

- Cycling of carbon, hydrogen, nitrogen, and oxygen between soil organic matter and the microbial biomass.

- Microbe-Mediated polymerization of the cycled organic compounds.

The investigation of molecular structure in humic substances is a difficult area of current research. Although it is not possible to describe the molecular configuration of humic substances in any but the most general terms, these molecules contain hydrogen ions that dissociate in fresh water to form molecules that bear a net negative charge. These negatively charged sites can interact with toxic metal ions and effectively remove them from further interaction with the environment.

Much of the molecular framework of soil organic matter, however, is not electrically charged. The uncharged portions of humic substances can react with synthetic organic compounds such as pesticides, fertilizers, solid and liquid waste materials, and their degradation products. Humus, either as a separate solid phase or as a coating on mineral surfaces, can immobilize these compounds and, in some instances, detoxify them significantly.

Biological Phenomena

Fertile soils are biological environments teeming with life on all size scales, from microfauna (with body widths less than 0.1 mm [0.004 inch]) to mesofauna (up to 2 mm [0.08 inch] wide) and macrofauna (up to 20 mm [0.8 inch] wide). The most numerous soil organisms are the unicellular microfauna: 1 kilogram (2.2 pounds) of soil may contain 500 billion bacteria, 10 billion actinomycetes (filamentous bacteria, some of which produce antibiotics), and nearly 1 billion fungi. The multicellular animal population can approach 500 million in a kilogram of soil, with microscopic nematodes (roundworms) the most abundant. Mites and springtails, which are categorized as mesofauna, are the next most prevalent. Earthworms, millipedes, centipedes, and insects make up most of the rest of the larger soil animal species. Plant roots also make a significant contribution to the biomass—the combined root length from a single plant can exceed 600 km (373 miles) in the top metre of a soil profile.

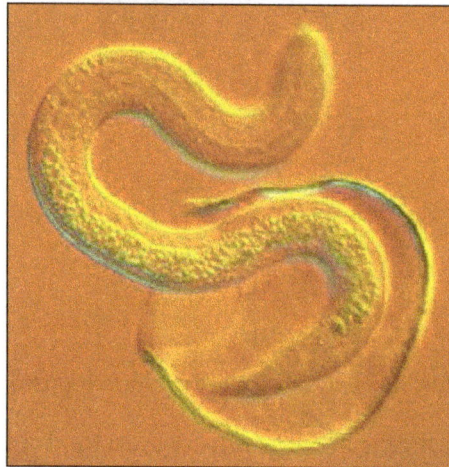

Roundworm (nematode) hatching. Nematodes feed on bacteria, fungi, and small animal forms in the soil.

The soil flora and fauna play an important role in soil development. Microbiological activity in the rooting zone of soils is important to soil acidity and to the cycling of nutrients. Aerobic and anaerobic (oxygen-depleted) microniches support microbes that determine the rate of the production of carbon dioxide (CO_2) from organic matter or of nitrate (NO_3^-) from molecular nitrogen (N_2).

The carbon and nitrogen cycles are two important microbe-mediated cycles. It is worth pointing out how they illustrate the complex, integrated nature of a soil's physical, chemical, and biological behaviour. Soil peds and pore spaces provide microniches for the action of carbon and nitrogen-cycling organisms, soil humus provides the nutrient reservoirs, and soil biomass provides the chemical pathways for cycling. The carbon in dead biomass is converted to CO_2 by aerobic microorganisms and to organic acids or alcohols by anaerobic microorganisms. Under highly anaerobic conditions, methane (CH_4) is produced by bacteria. The CO_2 produced can be used by photosynthetic microorganisms or by higher plants to create new biomass and thus initiate the carbon cycle again.

One of the primary pathways for the exchange of carbon dioxide (CO_2) takes place between the atmosphere and the oceans; there a fraction of the CO_2 combines with water, forming carbonic acid (H_2CO_3) that subsequently loses hydrogen ions (H^+) to form bicarbonate (HCO_3^-) and carbonate (CO_3^{2-}) ions. Mollusk shells or mineral precipitates that form by the reaction of calcium or other metal ions with carbonate may become buried in geologic strata and eventually release CO_2 through volcanic outgassing. Carbon dioxide also exchanges through photosynthesis in plants and through respiration in animals. Dead and decaying organic matter may ferment and release CO_2 or methane (CH_4) or may be incorporated into sedimentary rock, where it is converted to fossil fuels. Burning of hydrocarbon fuels returns CO_2 and water (H_2O) to the atmosphere. The biological and anthropogenic pathways are much faster than the geochemical pathways and, consequently, have a greater impact on the composition and temperature of the atmosphere.

The nitrogen (N) bound into proteins in dead biomass is consumed by microorganisms and converted into ammonium ions (NH_4^+) that can be directly absorbed by plant roots (for example, lowland rice). The ammonium ions are usually converted to nitrite ions (NO_2^-) by Nitrosomonas bacteria, followed by a second conversion to nitrate (NO_3^-) by Nitrobacter bacteria. This very mobile form of nitrogen is that most commonly absorbed by plant roots, as well as by microorganisms in soil. To close the nitrogen cycle, nitrogen gas in the atmosphere is converted to biomass nitrogen by Rhizobium bacteria living in the root tissues of legumes (e.g., alfalfa, peas, and beans) and leguminous trees (such as alder) and by cyanobacteria and Azotobacter bacteria.

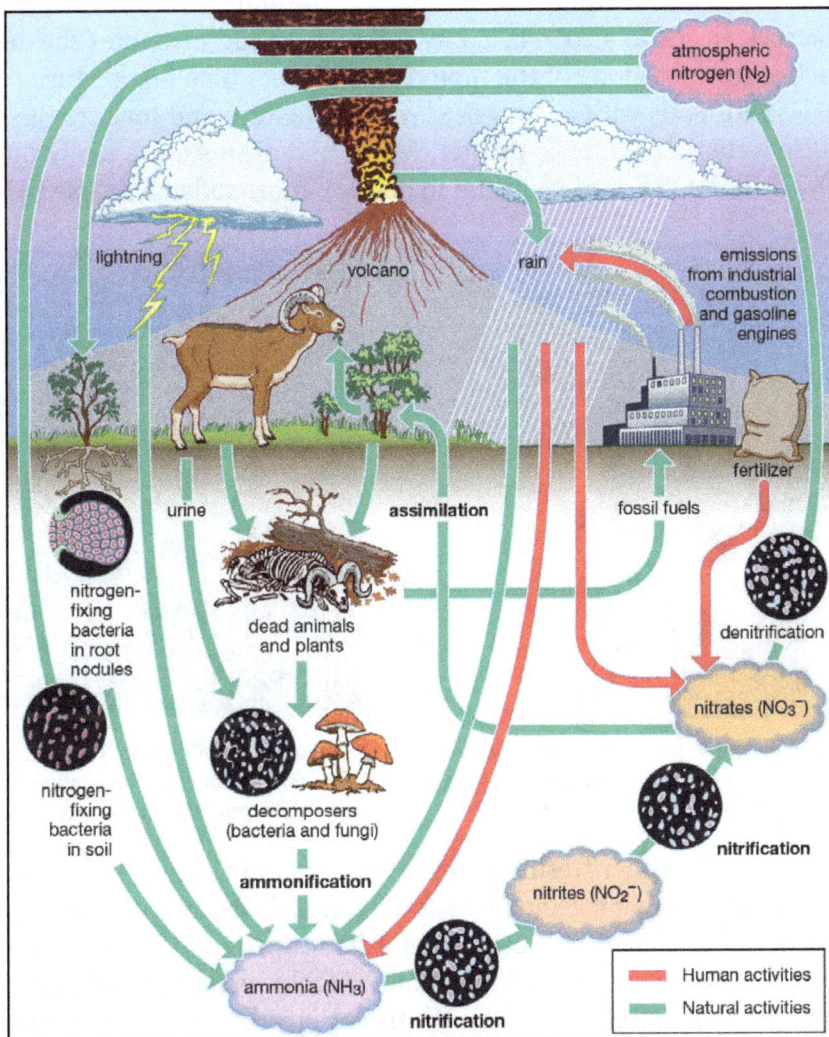

The nitrogen cycle.

Soil Formation

Soils evolve under the action of biological, climatic, geologic and topographic influences. The evolution of soils and their properties is called soil formation and pedologists have identified five fundamental soil formation processes that influence soil properties. These five "state factors" are: Parent material, Topography, Climate, Organisms and Time.

Parent Material

Parent material is the initial state of the solid matter making up a soil. It can consist of consolidated rocks, and it can also include unconsolidated deposits such as river alluvium, lake or marine sediments, glacial tills, loess (silt-sized, wind-deposited particles), volcanic ash, and organic matter (such as accumulations in swamps or bogs). Parent materials influence soil formation through their mineralogical composition, their texture, and their stratification (occurrence in layers). Dark-coloured ferromagnesian (iron- and magnesium-containing) rocks, for example, can produce soils with a high content of iron compounds and of clay minerals in the kaolin or smectite groups, whereas light-coloured siliceous (silica-containing) rocks tend to produce soils that are low in iron compounds and that contain clay minerals in the elite or vermiculite groups. The coarse texture of granitic rocks leads to a coarse, loamy soil texture and promotes the development of E horizons (the leached lower regions of the topmost soil layer). The fine texture of basaltic rocks, on the other hand, yields soils with a loam or clay-loam texture and hinders the development of E horizons. Because water percolates to greater depths and drains more easily through soils with coarse texture, clearly defined E horizons tend to develop more fully on coarse parent material.

Leptosol soil profile from Switzerland, showing a typically shallow surface horizon with little evidence of soil formation.

Fluvisol soil profile from South Africa, showing strata typical of sediments deposited from lakes, rivers, or oceans.

In theory, parent material is either freshly exposed solid matter (for example, volcanic ash immediately after ejection) or deep-lying geologic material that is isolated from atmospheric water and organisms. In practice, parent materials can be deposited continually by wind, water or volcanoes and can be altered from their initial, isolated state, thereby making identification difficult. If a single parent material can be established for an entire soil profile, the soil is termed monogenetic; otherwise, it is polygenetic. An example of polygenetic soils are soils that form on sedimentary rocks or unconsolidated water or wind-deposited materials. These so-called stratified parent materials can yield soils with intermixed geologic layering and soil horizons as occurs in south-eastern England, where soils forming atop chalk bedrock layers are themselves overlain by soil layers formed on both loess and clay materials that have been modified by dissolution of the chalk below:

Andosol soil profile from Italy, showing a dark-coloured surface
horizon derived from volcanic parent material.

Adjacent soils frequently exhibit different profile characteristics because of differing parent materials. These differing soil areas are called lithosequences, and they fall into two general types. Continuous lithosequences have parent materials whose properties vary gradually along a transect, the prototypical example being soils formed on loess deposits at increasing distances downwind from their alluvial source. Areas of such deposits in the central United States or China show systematic decreases in particle size and rate of deposition with increasing distance from the source. As a result, they also show increases in clay content and in the extent of profile development from weathering of the loess particles.

By contrast, discontinuous lithosequences arise from abrupt changes in parent material. A simple example might be one soil formed on schist (a silicate-containing metamorphic rock rich in mica) juxtaposed with a soil formed on serpentine (a ferromagnesian metamorphic rock rich in olivine). More subtle discontinuous lithosequences, such as those on glacial tills, show systematic variation of mineralogical composition or of texture in unconsolidated parent materials.

Topography

Topography, when considered as a soil-forming factor, includes the following: The geologic structural characteristics of elevation above mean sea level, aspect (the compass orientation of a landform), slope configuration (i.e., either convex or concave), and relative position on a slope (that is, from the toe to the summit). Topography influences the way the hydrologic cycle affects earth material, principally with respect to runoff processes and evapotranspiration. Precipitation may run off the land surface, causing soil erosion, or it may percolate into soil profiles and become part of subsurface runoff, which eventually makes its way into the stream system. Erosive runoff is most likely on a convex slope just below the summit, whereas lateral subsurface runoff tends to cause an accumulation of soluble or suspended matter near the toeslope. The conversion of precipitation into evapotranspiration is favoured by lower elevation and an equatorially facing aspect.

Adjacent soils that show differing profile characteristics reflecting the influence of local topography are called toposequences. As a general rule, soil profiles on the convex upper slopes in a toposequence are more shallow and have less distinct subsurface horizons than soils at the summit or on lower, concave-upward slopes. Organic matter content tends to increase from the summit down to the toeslope, as do clay content and the concentrations of soluble compounds.

Often the dominant effect of topography is on subsurface runoff (or drainage). In humid temperate regions, well-drained soil profiles near a summit can have thick E horizons (the leached layers) overlying well-developed clay-rich Bt horizons, while poorly drained profiles near a toeslope can have thick A horizons overlying extensive Bg horizons (lower layers whose pale colour signals stagnation under water-saturated conditions). In humid tropical regions with dry seasons, these profile characteristics give way to less distinct horizons, with accumulation of silica, manganese, and iron near the toeslope, whereas in semiarid regions soils near the toeslope have accumulations of the soluble salts sodium chloride or calcium sulfate.

These general conclusions are tempered by the fact that topography is susceptible to great changes over time. Soil erosion by water or wind removes A horizons and exposes B horizons to weathering. Major portions of entire soil profiles can move downslope suddenly by the combined action of water and gravity. Catastrophic natural events, such as volcanic eruptions, earthquakes, and devastating storms, can have obvious consequences for the instability of geomorphologic patterns.

Climate

The term climate in pedology refers to the characteristics of weather as they evolve over time scales longer than those necessary for soil properties to develop. These characteristics include precipitation, temperature, and storm patterns—both their averages and their variation.

Cryosol soil profile from Canada showing patterned deformations caused by cycles of freeze and thaw.

Climate influences soil formation primarily through effects of water and solar energy. Water is the solvent in which chemical reactions take place in the soil, and it is essential to the life cycles of soil organisms. Water is also the principal medium for the erosive or percolative transport of solid particles. The rates at which these water-mediated processes take place are controlled by the amount of energy available from the sun.

On a global scale, the integrated effects of climate can readily be seen along a transect from pole to Equator. As one proceeds from the pole to cool tundra or forested regions, polar desert soils give way to intensively leached soils such as the Podzols (Spodosols) that exhibit an eye-catching, ash-coloured E horizon indicative of humid, boreal climates. Farther into temperate zones, organic matter accumulates in soils as climates become warmer, and eventually lime (calcium carbonate) also begins to accumulate closer to the top of the soil profile as evapotranspiration increases. Arid subtropical climate then follows, with desert soils that are low in organic matter and enriched in soluble salts. As the climate again becomes humid close to the Equator, high temperature combines with high precipitation to create red and yellow tropical soils, whose colours reveal the prevalence of residual iron oxide minerals that are resistant to leaching losses because of their low solubility.

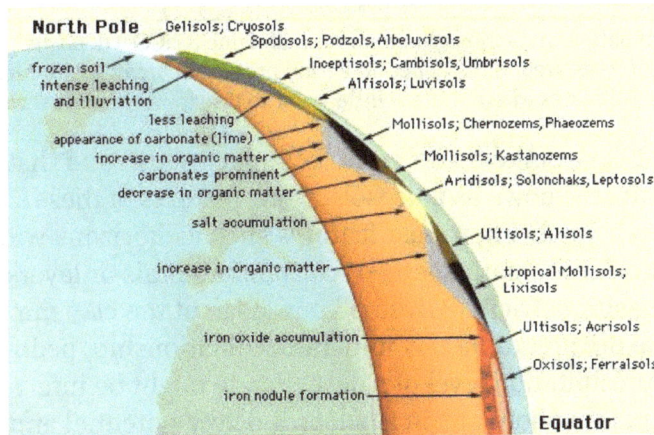

The effects of climate on soil worldwide A schematic cross section of the Earth illustrates the variation in soil types arising from differences in climate from the North Pole to the Equator.

On a continental scale, a transect taken across the central United States from east to west shows the effects of increasing evapotranspiration. First, soils that exhibit E horizons appear, followed by soils high in organic matter. These give way to soils with accumulations of lime and ultimately to desert soils with soluble salt efflorescence (powdery crust) near the surface.

On a regional scale, variations in climate also can influence soil properties significantly, resulting in a contiguous array of soils called a climosequence. One typical climosequence occurs along a 1,000-km (600-mile) north-south transect through the foothills of the Cascade and Sierra Nevada mountains in California. There soils that have formed on landscapes of similar topography vary continuously in their profile characteristics with variations in annual precipitation. Soils formed at the dry southern end of the transect are shallow and rocky, whereas those at the humid northern end show well-developed B horizons and reddish colour. Clay mineralogy in the upper 20 cm (8 inches) of these soils also responds to the increase in precipitation, shifting from the smectite group to the mixed vermiculite or illite group/kaolin group and finally to the kaolin group alone. These changes result primarily from increasing loss of silica and soluble metals as soil leaching extends deeper with increasing rainfall. In addition, soil acidity and organic matter content increase, while readily soluble forms of calcium (important to plant growth and soil aggregation) decrease, with increasing precipitation.

In principle, soil profile characteristics that are closely linked to climate can in turn be interpreted as climatic indicators. For example, a soil profile with two well-defined zones of lime accumulation, one shallow and one deep, may signal the existence of a past climate whose greater precipitation drove the lime layer deeper than the present climate is able to do.

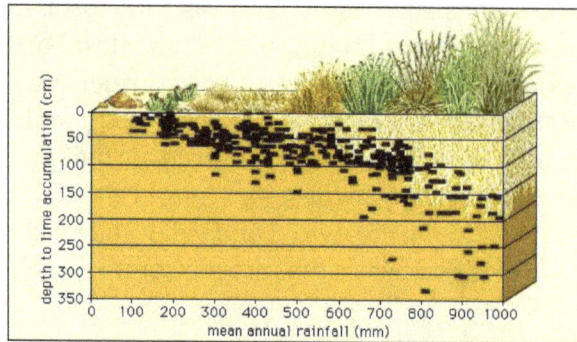

Depth to lime accumulation in relation to annual rainfall Lime (CaCO$_3$) deposits that can prevent the penetration of plant roots are found deeper in the soil profile in climates with higher mean annual rainfall than in climates where there is little water to transport the lime through the soil.

Soils that formed in past environments different from the present and that are preserved (at least partially) at greater depth are known as paleosols. Some features of these soils can serve as climatic indicators, the most reliable being robust features such as horizons with hardened accumulations of relatively insoluble iron, manganese, or calcium minerals or layers with accumulations of strongly aggregated clay-size particles. Given a knowledge of the clay mineral in a suspected paleosol, and assuming the precipitation-clay mineralogy relationship, pedologists might be able to infer past climate. The precipitation level of a past climate might be inferred from an observation of the depth of lime-containing horizons in a paleosol. These potential applications of climatic relationships must be evaluated carefully in order to distinguish the effects of previously weathered parent material from those of purely climatic influence.

Organisms

The development of soils can be significantly affected by vegetation, animal inhabitants, and human populations. Any array of contiguous soils influenced by local flora and fauna is termed a biosequence. To return to the climosequence along the Cascade and Sierra Nevada ranges the vegetation observed along this narrow foothill region varies from shrubs in the dry south to needle-leaved trees in the humid north, with extensive grasslands in between. In the middle of the precipitation range, transition zones occur in which small groves of needle-leaved trees are interspersed with grassland patches in an apparently random manner. These plant populations represent local flora largely selected by climate. The properties of the soils underlying these plants, however, exhibit differences that do not arise from climate, topography, or parent material but are an effect of the differing plant species. The soils under trees, for instance, are much more acidic and contain much less humus than those under grass, and nitrogen content is considerably greater in the grassland soil. These properties come directly from the type of litter produced by the two different kinds of vegetation.

An opportunity to examine biosequences is often presented by relatively young soils formed from an alluvial parent material. Soils of this kind lying beneath shrubs may be richer in humus and plant nutrients than similar soils found beneath needle-leaved trees. This variation results from differences in the cyclic processes of plant growth, litter production, and litter decay. Organic matter decomposers will feed on stored material in soil if litter production is low, whereas high litter production will permit soil stocks of organic matter to increase, leading to humus-rich A horizons as opposed to the leached E horizons found in soils that form under humid climatic conditions.

Human beings are also part of the biological influx that influences soil formation. Human influence can be as severe as wholesale removal or burial (by urbanization) of an entire soil profile or it can be as subtle as a gradual modification of organic matter by agriculture or of soil structure by irrigation. The chemical and physical properties of soils critical to the growth of crops often are affected significantly by cultural practices. Among the problems created for agriculture by cultural practices themselves are loss of arable land, erosion, the buildup of salinity, and the depletion of organic matter.

Anthrosol soil profile from The Netherlands, showing surface layers made homogeneous through human activity.

Time

The soil-forming factors of parent material and topography are largely site-related (attributes of the terrain), whereas those of climate and organisms are largely flux-related (inputs from the surroundings). Time as a soil-forming factor is neither a property of the terrain nor a source of external stimulus. It is instead an abstract variable whose significance is solely as a marker of the evolution of soil characteristics. The conceptual independence of time from its four companion factors means simply that soil evolution can occur while site attributes and external inputs remain essentially unchanged.

Certain soil profile features can be interpreted as indicators of the passage of time. (A series of soil profiles whose features differ only as a result of age constitutes a chronosequence.) One example of a time-related feature is the humus content of the A horizon, which, for soils less than 10,000 years old, increases continually at a rate dependent on parent material, vegetation, and climate. Typically, this rate of increase slows after about 10,000 years, plant nutrients begin to leach away, and a significant decline in humus content is observed for soils whose age approaches one million years. (Agricultural practices can interrupt this trend, causing a gradual drop in stored humus by 25 percent or more. Correlated with these changes are economically important soil properties, such as nutrient supply and retention capability, acidity, and aeration.).

The accumulation of clay and lime in soil profiles as a result of their translocation downward is also an indication of aging. For example, older soils that have formed on calcium-containing loess deposits have better-developed E and Bt horizons (as well as thinner A horizons) than younger soils forming on these deposits. Similarly, soils in a chronosequence developed on alluvium can exhibit a clayey hardpan after 100,000 years or so. Soils also tend to redden in colour as they age, irrespective of climatic conditions, reflecting the persistence of poorly crystalline or crystalline oxide minerals containing Fe^{3+}. Indeed, the dominant mineralogy of the clay-size particles in soils is itself a reliable indicator of soil age. Any particular sequence of predominant clay mineralogy found in a soil is known collectively as the set of Jackson-Sherman weathering stages. Each downward increment through the table corresponds to increasing mineral residence time, both among and within the three principal stages (early, intermediate, and advanced).

Jackson-Sherman Soil Weathering Stages

	Characteristic minerals in soil clay fraction	Characteristic chemical and physical conditions of soil	Characteristic soil profile features
Early stage	Gypsum, carbonates, olivine/pyroxene/amphibole, Fe^{2+}-bearing micas, and feldspars.	Low water and humus content, limited leaching, reducing environments, and limited time for weathering.	Minimally weathered soils all over the world, though mainly in arid regions where low rainfall keeps weathering to a minimum.
Intermediate stage	Quartz, mica/illite, vermiculite/chlorite, and smectite.	Retention of Na, K, Ca, Mg, Fe^{2+}, and silica; alkalinity and ineffective leaching; igneous rock rich in Ca, Mg, and Fe^{2+} but no Fe^{2+} oxides; easily hydrolyzed silicates; and transport of silica into the weathering zone.	Soils of temperate regions developed under grass or trees—i.e., the major agricultural soils of the world.
Advanced stage	Kaolin, aluminum and iron oxides, and titanium oxides.	Removal of Na, K, Ca, Mg, Fe^{2+}, and silica; effective leaching by fresh water; low pH; and dispersion of silica.	Intensely weathered soils of the humid tropics, frequently characterized by acidity and low fertility.

The early stage of weathering is recognized through the dominance of sulfates, carbonates, and primary silicates, other than quartz and muscovite, in the soil clay fraction. These minerals can survive only if soils remain very dry, very cold or very wet most of the time—that is, if they have limited exposure to water, air or solar energy. The intermediate stage features quartz, muscovite, and secondary aluminosilicates prominently in the clay fraction. These minerals can survive under leaching conditions that do not deplete silica and metals, such as calcium or magnesium, and that do not result in the complete oxidation of Fe^{2+}, which is then incorporated into illite and smectite clays. The advanced stage, on the other hand, is associated with intensive leaching and strong oxidizing conditions, such that only oxides of aluminum, Fe^{3+}, and titanium remain. Kaolin will be a dominant clay mineral group only if the removal of silica by leaching is not complete or if there is an encroachment of silica-rich waters—as can occur, for example, when water percolating through a soil profile at the upper part of a toposequence moves downslope into a soil profile at the lower part.

Implicit is the conclusion that more time is required to form soils featuring the more persistent minerals in the clay fraction. This conclusion is borne out by careful field studies worldwide in which the rate of soil horizon formation is determined. Soils whose clay mineralogy falls in the early stage require less than a decade to develop a centimetre (0.4 inch) of horizon thickness. Soils with dominant clay minerals in the intermediate stage do this in less than a century, whereas soils with dominant clay minerals in the advanced stage need several hundred years to form a centimetre of solum.

Soil Classification

The two principal systems of soil classification in use today are the soil order system of the U.S. Soil Taxonomy and the soil group system, published as the World Reference Base for Soil Resources, developed by the Food and Agriculture Organization (FAO) of the United Nations. Both of these systems are morphogenetic, in that they use structural properties as the basis of classification while also drawing on the five factors of soil formation described in the previous topic in choosing which properties to emphasize.

Central to both systems is the notion of diagnostic horizons, well-defined soil layers whose structure and origin may be correlated to soil-forming processes and can be used to distinguish among soil units at the highest level of classification. Diagnostic horizons may be found very near the land surface (epipedons) or deep in the soil profile (subsurface horizons); they need not correspond to the horizon letter designations.

Primary Diagnostic Horizons of Soil

	U.S. Soil Taxonomy	Defining features	FAO soil group system
Epipedons	Histic	Thick organic layer	Histic
	Mollic	Thick, dark, neutral to alkaline	Mollic
	Ochric	Pale or thin	Ochric
	Umbric	Thick, dark, acidic	Umbric
Subsurface horizons	Argillic	Clay mineral deposition	Argic
	Cambic	In situ mineral weathering only	Cambic
	Oxic	Highly weathered; aluminum oxide, iron oxide, and kaolin clay deposition	Ferralic
	Spodic	Aluminum oxide, iron oxide, and humus deposition	Spodic

The existence of a diagnostic horizon in a soil profile often is sufficient to indicate its taxonomic class at the level of order (U.S.) or group (FAO). For example, soil profiles with mollic epipedons are in the Mollisol order of the U.S. Soil Taxonomy. Alternatively, mollic A horizons occur distinctively in the FAO soil groups whose properties are conditioned by a steppe environment (that is, Chernozem, Kastanozem, and Phaeozem). The U.S. and FAO names both denote soils that have formed in plains under grassland vegetation, whose extensive root growth leads to a high content of humus in the A horizon. Often, however, the correspondence between the two taxonomic systems is not as close as in this example, a point quite evident when soil maps of the United States based on the U.S. and FAO taxonomies are compared.

Soils in Ecosystems

An ecosystem is a collection of organisms and the local environment with which they interact. For the soil scientist studying microbiological processes, ecosystem boundaries may enclose a single soil horizon or a soil profile. When nutrient cycling or the effects of management practices on soils are being considered, the ecosystem may be as large as an entire plant community and soil polypedon system.

Carbon and Nitrogen Cycles

Soils are dynamic, open habitats that provide plants with physical support, water, nutrients and air for growth. Soils also sustain an enormous population of microorganisms such as bacteria and fungi that recycle chemical elements, notably carbon and nitrogen, as well as, elements that are toxic. The carbon and nitrogen cycles are important natural processes that involve the uptake of nutrients from soil, the return of organic matter to the soil by tissue aging and death, the decomposition of organic matter by soil microbes (during which nutrients or toxins may be cycled within the microbial community), and the release of nutrients into soil for uptake once again. These cycles are closely linked to the hydrologic cycle, since water functions as the primary medium for chemical transport.

Nitrogen (N), one of the major nutrients, originates in the atmosphere. It is transformed and transported through the ecosystem by the water cycle and biological processes. This nutrient enters the biosphere primarily as wet deposition to the soil surface (throughfall), where plants, microbial decomposers, or nitrifiers (microbes that convert ammonium [NH_4^+] to nitrate [NO_3^-]) compete for it. This competition plays a major role in determining the extent to which incoming nitrogen will be retained within an ecosystem.

Carbon (C) also enters the ecosystem from the atmosphere—in the form of carbon dioxide (CO_2)—and is taken up by plants and converted into biomass. Organic matter in the soil in the form of humus and other biomass contains about three times as much carbon as does land vegetation. Soils of arid and semiarid regions also store carbon in inorganic chemical forms, primarily as calcium carbonate ($CaCO_3$). These pools of carbon are important components of the global carbon cycle because of their location near the land surface, where they are subject to erosion and decomposition. Each year, soils release 4–5 percent of their carbon to the atmosphere by the transformation of organic matter into CO_2 gas, a process termed soil respiration. This amount of CO_2 is more than 10 times larger than that currently produced from the burning of fossil fuels (coal and petroleum), but it is returned to the soil as organic matter by the production of biomass.

A large portion of the soil carbon pool is susceptible to loss as a result of human activities. Land-use changes associated with agriculture can disrupt the natural balance between the production of carbon-containing biomass and the release of carbon by soil respiration. One estimate suggests that this imbalance alone results in an annual net release of CO_2 to the atmosphere from agricultural soils equal to about 20 percent of the current annual release of CO_2 from the burning of fossil fuels. Agricultural practices in temperate zones, for example, can result in a decline of soil organic matter that ranges from 20 to 40 percent of the original content after about 50 years of cultivation. Although a portion of this loss can be attributed to soil erosion, the majority is from an increased flux of carbon to the atmosphere as CO_2. The draining of peatlands may cause similarly large losses in soil carbon storage.

Soils and Global Change

Soils and climate have always been closely related. The predicted temperature increases due to global warming and the consequent change in rainfall patterns are expected to have a substantial impact on both soils and demographics. This anticipated climatic change is thought to be driven by the greenhouse effect—an increase in levels of certain trace gases in the atmosphere such as carbon dioxide (CO_2), methane (CH_4), and nitrous oxide (N_2O). The conversion of land to agriculture, especially in the humid tropics, is an important contribution to greenhouse gas emissions. Some computer models predict that CH_4 and N_2O emissions will also be very important in future global change. About 70 percent of the CH_4 and 90 percent of the N_2O in the atmosphere are derived from soil processes. But soils can also function as repositories for these gases, and it is important to appreciate the complexity of the source-repository relationship. For example, the application of nitrogen-containing fertilizers reduces the ability of the soil to process CH_4. Even the amount of nitrogen introduced into soil from acid rain on forests is sufficient to produce this effect. However, the extent of net emissions of CH_4 and N_2O and the microbial trade-off between the two gases are undetermined at the global scale.

Perhaps the most notable and pervasive role of soils in global change phenomena is the regulation of the CO_2 budget. Carbon that is stored in terrestrial plants mainly through photosynthesis is called net primary production or NPP and is the dominant source of food, fuel, fibre, and feed for the entire population of Earth. Approximately 55 billion metric tons (61 billion tons) of carbon are stored in this way each year worldwide, most of it in forests. About 800 million hectares (20 billion acres) of forestland have been lost since the dawn of civilization; this translates to about 6 billion metric tons of carbon per year less NPP than before land was cleared for agriculture and commerce. This estimated decrease in carbon storage can be compared to the 5–6 billion metric tons of carbon currently released per year by fossil fuel burning. One is left with the sobering conclusion that reforestation of the entire planet to primordial levels would have only a temporary counterbalancing effect on carbon release to the atmosphere from human consumption of natural resources.

Carbon in terrestrial biomass that is not used directly becomes carbon in litter (about 25 billion metric tons of carbon annually) and is eventually incorporated into soil humus. Soil respiration currently releases an average of 68 billion metric tons of this carbon back into the atmosphere. The natural cycling of carbon is directly and indirectly affected by land-use changes through deforestation, reforestation, wood products decomposition, and abandonment of agricultural land. The current estimate of carbon loss from all these changes averages about 1.7 billion metric tons per year worldwide, or about one-third the current loss from fossil fuel burning. This figure could as much as double in the first half of the 21st century if the rate of deforestation is not controlled. Reforestation, on the other hand, could actually reduce the current carbon loss by up to 10 percent without exorbitant demands on management practices.

SOIL SCIENCE

Soil science is the study of soil as a natural resource on the surface of the Earth including soil formation, classification and mapping; physical, chemical, biological, and fertility properties of soils; and these properties in relation to the use and management of soils.

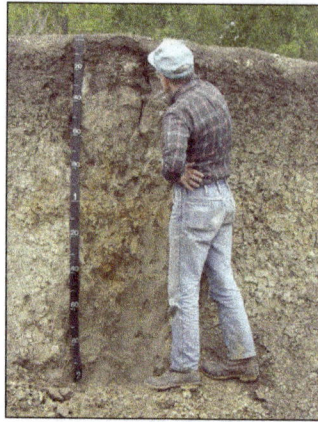

A soil scientist examines horizons within the soil profile.

Sometimes terms which refer to branches of soil science, such as pedology (formation, chemistry, morphology, and classification of soil) and edaphology (how soils interact with living things, especially plants), are used as if synonymous with soil science. The diversity of names associated with this discipline is related to the various associations concerned. Indeed, engineers, agronomists, chemists, geologists, physical geographers, ecologists, biologists, microbiologists, silviculturists, sanitarians, archaeologists, and specialists in regional planning, all contribute to further knowledge of soils and the advancement of the soil sciences.

Soil scientists have raised concerns about how to preserve soil and arable land in a world with a growing population, possible future water crisis, increasing per capita food consumption and land degradation.

Fields of Study

Soil occupies the pedosphere, one of Earth's spheres that the geosciences use to organize the Earth conceptually. This is the conceptual perspective of pedology and edaphology, the two main branches of soil science. Pedology is the study of soil in its natural setting. Edaphology is the study of soil in relation to soil-dependent uses. Both branches apply a combination of soil physics, soil chemistry, and soil biology. Due to the numerous interactions between the biosphere, atmosphere and hydrosphere that are hosted within the pedosphere, more integrated, less soil-centric concepts are also valuable. Many concepts essential to understanding soil come from individuals not identifiable strictly as soil scientists. This highlights the interdisciplinary nature of soil concepts.

Research

Dependence on and curiosity about soil, exploring the diversity and dynamics of this resource continues to yield fresh discoveries and insights. New avenues of soil research are compelled by a need to understand soil in the context of climate change, greenhouse gases, and carbon sequestration. Interest in maintaining the planet's biodiversity and in exploring past cultures has also stimulated renewed interest in achieving a more refined understanding of soil.

Mapping

Most empirical knowledge of soil in nature comes from soil survey efforts. Soil survey or soil mapping, is the process of determining the soil types or other properties of the soil cover over a landscape, and mapping them for others to understand and use. It relies heavily on distinguishing the individual influences of the five classic soil forming factors. This effort draws upon geomorphology, physical geography, and analysis of vegetation and land-use patterns. Primary data for the soil survey are acquired by field sampling and supported by remote sensing.

Classification

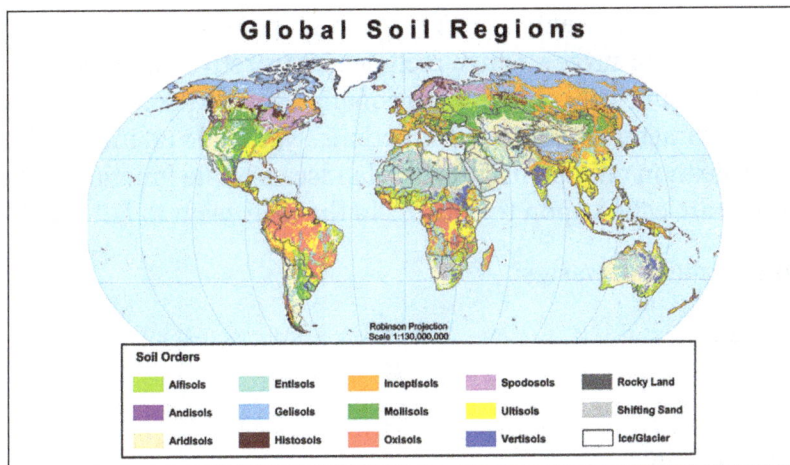

Map of global soil regions from the USDA.

As of 2006, The World Reference Base for Soil Resources, via its Land & Water Development division, is the pre-eminent soil classification system. It replaces the previous FAO soil classification.

The WRB borrows from modern soil classification concepts, including USDA soil taxonomy. The classification is based mainly on soil morphology as an expression pedogenesis. A major difference with USDA soil taxonomy is that soil climate is not part of the system, except insofar as climate influences soil profile characteristics.

Many other classification schemes exist, including vernacular systems. The structure in vernacular systems are either nominal, giving unique names to soils or landscapes or descriptive, naming soils by their characteristics such as red, hot, fat, or sandy. Soils are distinguished by obvious characteristics, such as physical appearance (e.g., color, texture, landscape position), performance (e.g., production capability, flooding), and accompanying vegetation. A vernacular distinction familiar to many is classifying texture as heavy or light. Light soil content and better structure, take less effort to turn and cultivate. Contrary to popular belief, light soils do not weigh less than heavy soils on an air dry basis nor do they have more porosity.

Areas of Practice

Academically, soil scientists tend to be drawn to one of five areas of specialization: microbiology, pedology, edaphology, physics, or chemistry. Yet the work specifics are very much dictated by the challenges facing our civilization's desire to sustain the land that supports it, and the distinctions between the sub-disciplines of soil science often blur in the process. Soil science professionals commonly stay current in soil chemistry, soil physics, soil microbiology, pedology, and applied soil science in related disciplines.

One interesting effort drawing in soil scientists in the USA as of 2004 is the Soil Quality Initiative. Central to the Soil Quality Initiative is developing indices of soil health and then monitoring them in a way that gives us long term (decade-to-decade) feedback on our performance as stewards of the planet. The effort includes understanding the functions of soil microbiotic crusts and exploring the potential to sequester atmospheric carbon in soil organic matter. The concept of soil quality, however, has not been without its share of controversy and criticism, including critiques by Nobel Laureate Norman Borlaug and World Food Prize Winner Pedro Sanchez.

A more traditional role for soil scientists has been to map soils. Most every area in the United States now has a published soil survey, which includes interpretive tables as to how soil properties support or limit activities and uses. An internationally accepted soil taxonomy allows uniform communication of soil characteristics and soil functions. National and international soil survey efforts have given the profession unique insights into landscape scale functions. The landscape functions that soil scientists are called upon to address in the field seem to fall roughly into six areas.

- Land-based treatment of wastes:
  - Septic system.
  - Manure.
  - Municipal biosolids.
  - Food and fiber processing waste.
- Identification and protection of environmentally critical areas:
  - Sensitive and unstable soils.
  - Wetlands.
  - Unique soil situations that support valuable habitat, and ecosystem diversity.
- Management for optimum land productivity:
  - Silviculture.
  - Agronomy:
    - Nutrient management.
    - Water management.
  - Native vegetation.
  - Grazing.
- Management for optimum water quality:
  - Stormwater management.
  - Sediment and erosion control.

- Remediation and restoration of damaged lands:
  - Mine reclamation.
  - Flood and storm damage.
  - Contamination.
- Sustainability of desired uses:
  - Soil conservation.

There are also practical applications of soil science that might not be apparent from looking at a published soil survey.

- Radiometric dating: specifically a knowledge of local pedology is used to date prior activity at the site:
  - Stratification (archeology) where soil formation processes and preservative qualities can inform the study of archaeological sites.
  - Geological phenomena:
    - Landslides.
    - Active faults.
- Altering soils to achieve new uses:
  - Vitrification to contain radioactive wastes.
  - Enhancing soil microbial capabilities in degrading contaminants (bioremediation).
  - Carbon sequestration.
  - Environmental soil science.
- Pedology:
  - Soil genesis.
  - Pedometrics.
  - Soil morphology:
    - Soil micromorphology.
  - Soil classification:
    - USDA soil taxonomy.
- Soil biology:
  - Soil microbiology.

- Soil chemistry:
  - Soil biochemistry.
  - Soil mineralogy.
- Soil physics:
  - Pedotransfer function.
  - Soil mechanics and engineering.
- Soil hydrology, hydropedology.

Fields of Application in Soil Science

- Climate change.
- Ecosystem studies.
- Pedotransfer function.
- Soil fertility/Nutrient management.
- Soil management.
- Soil survey.
- Standard methods of analysis.
- Watershed and wetland studies.

Related Disciplines

- Agricultural sciences:
  - Agricultural soil science.
  - Agrophysics science.
  - Irrigation management.
- Anthropology:
  - Archaeological stratigraphy.
- Environmental science:
  - Landscape ecology.
- Physical geography:
  - Geomorphology.

- Geology:
  - Biogeochemistry.
  - Geomicrobiology.
- Hydrology:
  - Hydrogeology.
- Waste management.
- Wetland science.

Depression Storage Capacity

Depression storage capacity, in soil science, is the ability of a particular area of land to retain water in its pits and depressions, thus preventing it from flowing. Depression storage capacity, along with infiltration capacity, is one of the main factors involved in Horton overland flow, whereby water volume surpasses both infiltration and depression storage capacity and begins to flow horizontally across land, possibly leading to flooding and soil erosion. The study of land's depression storage capacity is important in the fields of geology, ecology, and especially hydrology.

Types of Soil

There are various types of soils which have different origins, compositions, functions and different uses. These can be categorized into sandy soil, silt soil, clay soil and loam soil. The topics elaborated in this chapter will help in gaining a better perspective about these various types of soil.

Soil is a natural resource that can be categorised into different types, each with distinct characteristics that provide growing benefits and limitations. There are various types of soil that undergo diverse environmental pressures. The soil is mainly classified by its texture, proportions and different forms of organic and mineral compositions.

The soil is basically classified into four types:

- Sandy soil.

- Silt Soil.

- Clay Soil.

- Loamy Soil.

Sandy Soil

The first type of soil is sand. It consists of small particles of weathered rock. Sandy soils are one of the poorest types of soil for growing plants because it has very low nutrients and poor in holding water, which makes it hard for the plant's roots to absorb water. This type of soil is very good for the drainage system. Sandy soil is usually formed by the breakdown or fragmentation of rocks like granite, limestone, and quartz.

Sandy Soil - It has the largest particle among the three.

Silt Soil

Silt, which is known to have much smaller particles compared to the sandy soil and is made up of rock and other mineral particles which are smaller than sand and larger than clay. It is the smooth and quite fine quality of the soil that holds water better than sand. Silt is easily transported by moving currents and it is mainly found near the river, lake, and other water bodies. The slit soil is more fertile compared to the other three types of soil. Therefore it is also used in agricultural practices to improve soil fertility.

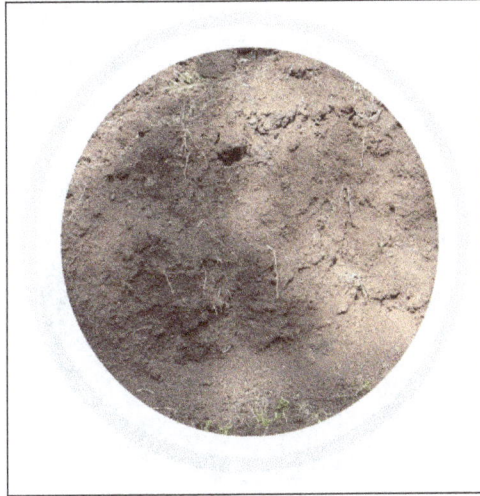

Silty Soil - Finer particles when compared to sand.

Clay Soil

Clay is the smallest particles amongst the other two types of soil. The particles in this soil are tightly packed together with each other with very little or no airspace. This soil has very good water storage qualities and making hard for moisture and air to penetrate into it. It is very sticky to the touch when wet, but smooth when dried. Clay is the densest and heaviest type of soil which do not drain well or provide space for plant roots to flourish.

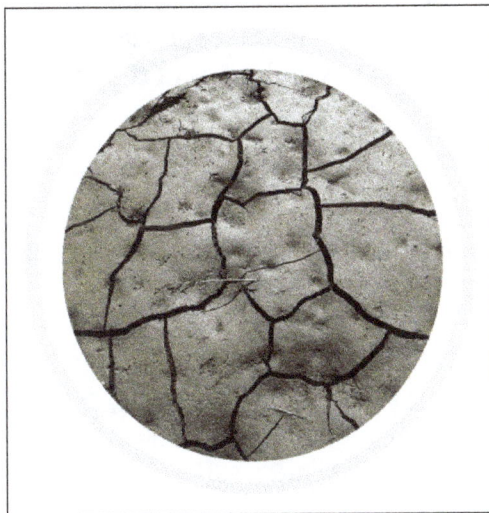

Clay Soil - It has the smallest particle among the three.

Loamy Soil

Loam is the fourth type of soil. It is a combination of sand, silt, and clay such that the beneficial properties from each is included. For instance, it has the ability to retain moisture and nutrients, hence, it is more suitable for farming. This soil is also referred to as an agricultural soil as it includes an equilibrium of all three types of soil materials being sandy, clay, and silt and it also happens to have hummus. Apart from these, it also has higher calcium and pH levels because of its inorganic origins.

Loamy Soil - Also known as agricultural soil.

SANDY SOIL

When the percentage of sand is high in a specific soil than it is called sandy soil. It has the largest particle among different soil particles. Sandy soil is also known as "Light soil". Generally, sandy soil is composed of- 35% sand and less than 15% silt and clay. Primarily sand is the small pieces of eroded rocks with a gritty texture. In sandy soil, most of the soil particulars are bigger than 2mm in diameter.

Uses of Sandy Soil

The uses of sandy soil in different sectors are numerous. Followings are the uses of sandy soil:

- Agricultural need,
- Easy drainage,
- Construction,
- Foundation,
- Beauty,

- Frictional properties,

- Low settlements,

- Changing pH,

- Others.

Agricultural Need

Sandy soil is usually dry, nutrient and fast draining. It is used for plowing, planting and cultivating. The useful vegetables like potatoes, grams, tomatoes, etc require a minimum percentage of soil for a specific period. The percentage varies from vegetable to vegetable. Sandy soil also provides a good ground for farmers to collect falling nuts.

Easy Drainage

Sandy soil has great drainage properties. It drains easily and quickly. It is used to improve soil drainage. The interesting feature is in a land full of sandy soil, work can be done right after a rain even if it's heavy without any difficulty. It filters water in big deposits through which water is shred and recollected through the channels at the bottom. Sandy soil also retains flower water like soaking the water.

Construction

Sandy soil doesn't get sticky. It is cohesionless. It has a light and loose structure. That's why it can be easily used for construction purpose. Sandy soil can be a great aggregate for concrete. Also, it can be used as a construction material of mortar aside cement. Sandy soil is also used for the erection of exterior rendering materials. It is used because of its chemical resistance. Sandy soil also can be used the best as filling sand.

Foundation

Sandy soil provides a base to foundations on swampy ground. Densification below foundation is not required as the soil is naturally in a dense state. Sandy soil provides a reliable foundation for which the structure that relies on this soil has a greater degree of reliance than these built-in other soils. Sandy soil is maneuvered for the improvement of ground with the use of soil replacement method. It is used to replace soft clays of foundation to improve the bearing capacity of the soil.

Beauty

Sandy soil starts from the almost sandy surface like beaches. It increases the beauty of the beach. Also, it's used in gardening and kids playgrounds for safety by providing a soothing context.

Frictional Properties

Sandy soil has very good frictional properties. The frictional properties are used in the construction of reinforced soil structure with geosynthetics reinforcement.

Low Settlements

Sandy soils have low settlements as it does not undergo consolidation with time. Moreover, it has immediate settlements.

Changing pH

The pH level of sandy soil can easily change the pH level of soil like clay. The pH level of sandy soil is between 7.00 and 8.00.

Others

Sandy oil is used to reduce the velocity of the water. It also helps to percolate the soil and raise the water table.

SILT SOIL

Silt

Silt is granular material of a size between sand and clay, whose mineral origin is quartz and feldspar. Silt may occur as a soil (often mixed with sand or clay) or as sediment mixed in suspension with water (also known as a suspended load) and soil in a body of water such as a river. It may also exist as soil deposited at the bottom of a water body, like mudflows from landslides. Silt has a moderate specific areawith a typically non-sticky, plastic feel. Silt usually has a floury feel when dry, and a slippery feel when wet. Silt can be visually observed with a hand lens, exhibiting a sparkly appearance. It also can be felt by the tongue as granular when placed on the front teeth (even when mixed with clay particles).

Silt is created by a variety of physical processes capable of splitting the generally sand-sized quartz crystals of primary rocks by exploiting deficiencies in their lattice. These involve chemical weathering of rock and regolith, and a number of physical weathering processes such as frost shattering and haloclasty. The main process is abrasion through transport, including fluvial comminution, aeolian attritionand glacial grinding. It is in semi-arid environments that substantial quantities of silt are produced. Silt is sometimes known as "rock flour" or "stone dust", especially when produced by glacial action. Mineralogically, silt is composed mainly of quartz and feldspar. Sedimentary rock composed mainly of silt is known as siltstone. Liquefaction created by a strong earthquake is silt suspended in water that is hydrodynamically forced up from below ground level.

Environmental Impacts

A silted lake located in Eichhorst, Germany.

Silt is easily transported in water or other liquid and is fine enough to be carried long distances by air in the form of dust. Thick deposits of silty material resulting from deposition by aeolian processes are often called loess. Silt and clay contribute to turbidity in water. Silt is transported by streams or by water currents in the ocean. When silt appears as a pollutant in water the phenomenon is known as siltation.

Silt, deposited by annual floods along the Nile River, created the rich, fertile soil that sustained the Ancient Egyptian civilization. Silt deposited by the Mississippi River throughout the 20th century has decreased due to a system of levees, contributing to the disappearance of protective wetland-sand barrier islands in the delta region surrounding New Orleans.

In southeast Bangladesh, in the Noakhali district, cross dams were built in the 1960s whereby silt gradually started forming new land called "chars". The district of Noakhali has gained more than 73 square kilometres (28 sq mi) of land in the past 50 years.

With Dutch funding, the Bangladeshi government began to help develop older chars in the late 1970s, and the effort has since become a multi-agency operation building roads, culverts, embankments, cyclone shelters, toilets and ponds, as well as distributing land to settlers. By fall 2010, the program will have allotted some 100 square kilometres (20,000 acres) to 21,000 families.

A main source of silt in urban rivers is disturbance of soil by construction activity. A main source in rural rivers is erosion from plowing of farm fields, clearcutting or slash and burn treatment of forests.

CLAY SOIL

Clay soil is defined as soil that comprised of very fine mineral particles and not much organic material. The resulting soil is quite sticky since there is not much space between the mineral particles, and it does not drain well at all. If you have noticed that water tends to puddle on the ground rather than soak in, it is likely your ground consists of clay. Soil that consists of over 50 percent clay particles is referred to as "heavy clay." To determine whether you have clay soil or not, you can do a simple soil test. If your soil sticks to shoes and garden tools like glue, forms big clods that aren't easy to separate, and crusts over and cracks in dry weather, you have clay.

Advantages of Clay Soil

Even clay soil has some good qualities. Clay, because of its density, retains moisture well. It also tends to be more nutrient-rich than other soil types. The reason for this is that the particles that make up clay soil are negatively charged, which means they attract and hold positively charged particles, such as calcium, potassium, and magnesium.

Disadvantages of Clay Soil

In addition to the drawbacks mentioned above, clay also has the following negative qualities:

- Slow draining.
- Slow to warm in the spring.
- Compacts easily, making it difficult for plant roots to grow.
- Tendency to heave in winter.
- Tendency to be alkaline in pH.

Properties of Clay Soil

Though different soils have a wide range of colors, textures and other distinguishing features, there are only three types of soil particles that geologists consider distinct. The quality of soil depends on the amount of sand, loam and clay that it contains, because soils with differing amounts of these particles often have very different characteristics. Soil with a large amount of clay is sometimes hard to work with, due to some of clay's characteristics.

Particle Size

Clay has the smallest particle size of any soil type, with individual particles being so small that they can only be viewed by an electron microscope. This allows a large quantity of clay particles to exist in a relatively small space, without the gaps that would normally be present between larger soil particles. This feature plays a large part in clay's smooth texture, because the individual particles are too small to create a rough surface in the clay.

Structure

Because of the small particle size of clay soils, the structure of clay-heavy soil tends to be very dense. The particles typically bond together, creating a mass of clay that can be hard for plant roots to penetrate. This density is responsible for clay-heavy soil being thicker and heavier than other soil types, and clay soil takes longer to warm up after periods of cold weather. This density also makes clay soils more resistant to erosion than sand or loam-based soils.

Organic Content

Clay contains very little organic material; you often need to add amendments if you wish to grow plants in clay-heavy soil. Without added organic material, clay-heavy soil typically lacks the nutrients and micronutrients essential for plant growth and photosynthesis. Mineral-heavy clay soils may be alkaline in nature, resulting in the need for additional amendments to balance the soil's pH before planting anything that prefers a neutral pH. It's important to test clay-heavy soil before planting to determine both the soil's pH and whether it lacks important nutrients such as nitrogen, phosphorus and potassium.

Permeablity and Water-holding Capacity

One of the problems with clay soil is its slow permeability resulting in a very large water-holding capacity. Because the soil particles are small and close together, it takes water much longer to move through clay soil than it does with other soil types. Clay particles then absorb this water, expanding as they do so and further slowing the flow of water through the soil. This not only prevents water from penetrating deep into the soil but can also damage plant roots as the soil particles expand.

Identifying Clay

There are several tests you can use to identify clay soils. If rubbed between your fingers, a sample of clay soil often feels slick and may stick to your fingers or leave streaks on your skin. Rubbed clay soil often takes on a shiny appearance as well, as opposed to the rough texture you would see with other soils. Clay soils do not crumble well, and a sample of clay can typically be stretched slightly without breaking. When wet, clay soils become slick and sticky; the soil may also allow water to pool briefly before absorption due to the slow permeation. Visually, clay soils seem solid with no clear particles, and may have a distinct red or brown color when compared to the surrounding soil.

LOAM SOIL

Loam soil is soil made up of sand particles, silt, and clay. The sand, silt and clay particles have a size of greater than 63 micrometers, greater than two micrometers and less than two micrometers, respectively. Additionally, loam soil is typically composed of these three components in the following weight ratio: 40% sand, 40% silt, and 20% clay. However, these are general proportions and may vary from one sample to another, which leads to the formation of different kinds of loam soils, such as sandy clay loam, silty loam, silty clay loam and others. The United States Department of Agriculture (USDA) usually classifies any soil that is not predominantly sand, clay or silt as loam. The soil feels crumbly and soft and is relatively easier to work regardless of the moisture level.

Characteristics of Loam Soil

The drainage of loam soil is usually affected by its composition.

Loam soil that has a higher amount of sand and low amounts of organic content tends to show greater drainage capabilities. However, because of the clay particles in the soil, loam tends to hold water relatively well. Loam with a higher content of organic matter tends to dry out faster. Loam soil with higher amounts of clay is also more compact, making it ideal for use in surfaces that require compacted sand, such as roads.

Loam often has a fair level of aeration. Consequently, it is ideal for the survival of organisms in the soil that are beneficial for plant life. However, loam that has higher amounts of clay is not that conducive for plant life or soil organisms. Farmers usually have problems dealing with loam soil that has higher concentrations of clay.

Loam is average in terms of maintaining nutrient levels in the soil. Sandy loam has fewer nutrients since it drains faster than loam that retains moisture. However, the ability of loam to hold nutrients can be greatly increased by adding compost.

Loam Soil and Farming

Given its ability to retain water and nutrients yet allow drainage, loam soil is excellent for agricultural use. The composition of the loam is less important since the soil only needs to have these two qualities. In some cases, a soil sample may satisfy the definition of loam but still be unsuitable for agriculture due to poor drainage or low nutrient retention or both. Some of the prominent agricultural regions in the world have loam soil.

Loam Soil and House Construction

In some parts of the world, loam soil has proven useful in house construction, such as the post and beam method. In some cases, builders use loam to construct a layer on the inside walls of a house to help manage air humidity. However, one of the oldest and most common uses of loam in construction is to mix it with straw and use the mixture to build walls.

References

- Types-of-soil, biology: byjus.com, Retrieved 30 March, 2019

- "Particle Size (618.43)". National Soil Survey Handbook Part 618 (42-55) Soil Properties and Qualities. United States Department of Agriculture - Natural Resource Conservation Service. Archived from the original on 2006-05-27. Retrieved 2006-05-31

- 213-sandy-soil-uses-of-sandy-soil, sand, civil-engineering-materials: civiltoday.com, Retrieved 1 April, 2019

- Lautridou, J. P.; Ozouf, J. C. (19 August 2016). "Experimental frost shattering". Progress in Physical Geography. 6 (2): 215–232. doi:10.1177/030913338200600202

- Understanding-and-improving-clay-soil-2539857: thespruce.com, Retrieved 2 May 2019

- Properties-clay-soil-71840: homeguides.sfgate.com, Retrieved 3 June, 2019

- What-is-loamy-soil: worldatlas.com, Retrieved 4 July, 2019

Fundamental Concepts of Soil Science

Some of the most important fundamental concepts of soil science are soil fertility, soil respiration, soil functions, soil water retention, soil genesis, soil morphology, soil ecology, soil biology, humus and soil horizon. This chapter has been carefully written to provide an easy understanding of these fundamental concepts of soil science.

SOIL FERTILITY

Soil fertility may be defined as the ability of soil to provide all essential plant nutrients in available forms and in a suitable balance whereas soil productivity is the resultant of several factors such as soil fertility, good soil management practices availability of water supply and suitable climate.

The soil is a natural medium for plant growth and it supplies nutrients to plants. Some soils areW productive and they support luxuriant growth of plants with very little human effort whereas others may be unproductive which support almost no useful plant life regardless of every human effort In order for soil to be productive, it must:

- Be easily tillable and fertile.

- Contain all essential elements in the forms readily available to plants in sufficient amount.

- Physically good to support plants and contain just the right amount of water and air for proper root growth. The soil must supply these essentials every day in the life of the plant.

Soil fertility and soil productivity appear to be synonymous but in soil science these two terms bear different meanings. Soil fertility may be defined as the ability of soil to provide all essential plant nutrients in available forms and in a suitable balance whereas soil productivity is the resultant of several factors such as soil fertility, good soil management practices availability of water supply and suitable climate.

A soil can be highly fertile, i.e., it has ready supply of nutrients in available form, yet it may not be highly productive. Water-logged soils may be highly fertile but may not produce good crop because of the un-favourable physical conditions.

A fertile soil may be highly saline or alkaline which may not be good for agriculture Sandy soil may be poor in fertility but with the use of fertilizers and water it may be made productive. Soil fertility thus denotes the status of plant nutrients in the soil whereas the soil productivity is the resultant of various factors influencing crop production. In fact there is no standard for either fertility or productivity because both depend upon the crops to be grown. Soil that is productive for potatoes may not necessarily be productive for certain other crops.

Plant Nutrients

Of the 90 or so chemical elements forming the earth's crust, 16 are known to be essential for plant growth and reproduction. Seven elements needed in good quantity (macro nutrients) are hydrogen, oxygen, nitrogen and carbon from air and water and phosphorus, potassium and calcium from mineral particles in the soil. The other 9 elements needed only in small amount (micronutrients) are magnesium, sulphur, boron, copper, iron, manganese, zinc, molybdenum and chlorine.

Jacob and Uexkull have listed cobalt and sodium as essential elements for plant growth. With the exception of hydrogen, carbon, and oxygen all other inorganic plant requirements are obtained directly or indirectly from the soil minerals, hence these elements are called mineral nutrients. Strictly speaking, nitrogen is not a mineral element but it has been included in the list because it can also be obtained by plants from soil.

The mineral elements are taken by plants from soils mostly in the form of ions. Plants obtain nutrients from the following four devices:

- From the soil solutions through roots.
- From exchangeable ions on the surface of clay and humus particles through roots.
- From readily decomposable minerals.
- Through the leaves.

The essentiality of an element is proved by the following criteria:

- The element may be considered essential if its exclusion from the nutrient medium inhibits or drastically reduces the growth and reproduction of plant.
- Acute deficiency of the element produces certain well defined disease symptoms which are not produced by the deficiency of any other element.
- Deficiency disease symptoms will disappear if the particular element is supplied before the living system has been damaged beyond repairs.

The capacity of soil to supply the essential elements is a fundamental edaphological problem In natural habitat the plant matter is returned to ground through decay and unless there is extensive leaching and percolation the inorganic nutrients become available again for new growth Many farm crops are removed from ground in to and the supply of essential elements thereby becomes depleted.

To replace these elements taken away from ground by crops, the manures or fertilizers are added. Unless it is done the crop will suffer from deficiencies in the essential elements. Deficiencies become reflected in the growth of plants in several ways; some may cause a reduction in yield as a result of poor plant growth and some may delay maturation of crop a function that may be very vital to crop yield in the places where the growing season is short.

The symptoms of mineral deficiencies may be dwarf, spotted, distorted, curled or wilted leaves or rotting of the centre of fruits. Different mineral nutrients have certain specific deficiency symptoms.

Soil Fertility Factors

Several factors are known to govern the fertility of soil. Some of the important factors are:

- As a result of cropping, a large amount of organic matter and soil minerals are removed and if the normal cycling of mineral elements is retarded, loss in soil fertility may result.

- Besides cropping, soil erosion and loss of water also causes tremendous loss of plant nutrients from the top soil.

Generally, water is lost through leaching, drainage, evapotranspiration and runoff.

The following adverse affects are observed due to water loss:

- Soil becomes very hard.

- The seed germination percentage decreases.

- The nutrients in the soil leach or evaporate.

- The root growth retards, so that plants become stunted and, as a result, the yield is reduced.

- Stomata become closed, as a result of which accumulation of gases or metabolic wastes increases in plant tissues leading to death of the plant.

- The activity of soil-micro organisms decreases:

  ○ Conversion of organic forms of nitrogen locked in humus into ammonia gas and nitrogen gas and leaching out of soluble nitrates and nitrites from surface soil greatly affect the fertility status of soil.

  ○ Like deficiency, the abundance of certain nutrient elements in soluble form may also be toxic and even the elements, say alkalies, essential for plant growth may be toxic if present in excess. Flowering plants do not grow in the soil containing more than 6 per cent NaCl and other salts. The elements are not equally toxic and the various species of plants differ in their susceptibility to different elements.

  ○ Toxic chemicals and pesticides in soil: Several agricultural chemicals being used for controlling various diseases and insect pests are highly toxic and their application adversely affects the soil micro flora and fauna. Prolonged persistence of these pesticides in soil is bound to lower the soil fertility both directly and indirectly.

  ○ Soil reaction: The soils may be alkaline or neutral or acidic in their reaction. Some plants find acid soil unsuitable for growth and other plants find alkaline ground un-favourable. pH value of soil solution determines the availability of certain plant nutrients and thus it has bearing on soil fertility problem. Increase in the acidity of the soil makes mineral salts more soluble in soil solution and thus salts may become available in concentrations that may be highly toxic or may damage plants growing in such soils. Janick et al. have demonstrated that high concentration of both iron and aluminium may damage plants growing in acid soils.

Maintenance of Soil Fertility

Soil fertility is the most important asset of a nation. Maintenance of soil fertility is an important aspect of agriculture. The soil fertility problem has been studied in many countries and scientists have brought to light several facts concerning soil fertility and its maintenance.

Soil Fertility is of Two Types

- Permanent fertility: It is derived from the soil itself. It can be improved, maintained or corrected by soil management practices.

- Temporary fertility: It is acquired by suitable soil management but the response of built up soil fertility is highly dependent on the degree of permanent fertility which is already there. Several methods are known for controlling the loss of soil fertility. Here only the important methods are discussed.

Application of Organic Manures and Chemical Fertilizers

Plants absorb water and minerals from the soil, which is essential for growth, flowering, crop yield, and other vital activities. Soil is a store house for organic and inorganic plant nutrients. Some soils are rich in organic and humus content and are considered to be fertile and more productive while others that are deficient in humus and minerals are less productive.

The soil is subjected to a continuous depletion of nutrients due to its continuous use by crops. This requires the addition of mineral resources. The various soil components are being removed by living organisms and are returned to the soil by death and decay of organisms. If the rate of removal or loss of minerals is greater than the rate of addition, the soil will naturally become less fertile. The minerals of the soils are lost due to crops, leaching or soil erosion.

The minerals are often removed from the top layer by rainwater. Cultivation of crops regularly, year after year, makes the soil less productive. In intensive cultivation there are little chances for the restoration of lost nutrients in the soil until they are supplied from outside. The leguminous plants, however, compensate the loss of nitrogenous compounds. Besides this, manure and fertilizers are to be supplemented to restore the fertility of the soil.

The deficiency of mineral nutrients in the soil either can be compensated through organic manures such as green manuring, compost etc. or it can be supplemented by the application of chemical fertilizers from outside sources.

Organic Manures

The organic content of the soil which is a good source of plant nutrients contributes most to the fertility of the soil.

Organic manures improve soil fertility in the following ways:

- They modify the physical properties as increase in granulation of the soil and increase in permeability and moisture holding capacity of soil.

- They provide food for soil microbes and thus enhance microbial activities.

- Decomposition products of organic manures help to bring mineral constituents of soil into solution.

- They improve physico-chemical properties of soil, such as cation exchange and buffering action.

Organic manures are of several kinds some of which are:

- Farmyard manures: Solid and liquid excreta as dung and urine of all farm animals are termed farmyard manures. They are ready made manures and contain nitrogen, phosphorus and potassium. The farmyard manures of different animals vary greatly in their composition but they are good for all types of soils and all the crops. Farmyard manure when collected in field in exposed condition for several months shows considerable loss of fertilizing value as upon decomposition a considerable amount of ammonia is lost by volatilization.

Therefore, it is important to keep manure protected from weather and manure preparation should be carried out in trenches of about a metre depth. When the trenches are filled with dung etc, the surface is covered with a cow dung-earth slurry. In about 3 months the manure becomes ready for use.

- Compost: Compost manure can be prepared from a variety of refuse materials, such as straw, sugar cane refuse, rice hulls, forest, litter, weeds, leaves, kitchen wastes. It is prepared in pits usually 6-8 inch long, 1½ to 2 inch wide and one metre deep. In the pits, 30 cm thick layer of plant residues moistened with dung, urine and water is formed and then a second layer of about 30 cm thickness of mixed refuse is spread over it and moistened with slurry. The operation is repeated until the heap rises to a height of about 50 cm above the ground level. The top is then covered with a thin layer of moist earth. After three months of decomposition the material is well mixed and again covered. After a couple of months the manure is ready for use. There are two types of composts:

  ◦ Farmyard compost which is obtained from animal excreta and plant residues.

  ◦ Town compost which is obtained by decomposition of kitchen wastes and garbage of towns and cities. Compost manures are rich in all plant nutrients.

- Green manures: Green manuring is the practice of growing, ploughing and mixing of green crops with soil to improve soil fertility and productivity. Its effects on soils are similar to those of farmyard manures. It is cheap and the best method to increase soil fertility as it can supplement farmyard and other organic manures without involving much cost. Green manures add nitrogen and organic matter to the soil for the improvement of crop yield.

Through green manuring mobilization of minerals, reduction of organic nutrient losses due to erosion, leaching and percolation, and improvement in physical, chemical and biological activities of the soil can be achieved. Green manuring also improves soil aeration and drainage conditions. For green manuring both leguminous and non-leguminous crops are used. In India, leguminous crops such as sannhemp (sanai), dhaincha, berseem, clover, Phaseolus mungo, cowpea, are generally used for green manuring.

- Sawdust: Sawdust can be used as bedding material to conserve animal urine or for making compost. It is a low fertilizing material but it is definitely richer than wheat straw in calcium.

- Sewage: In modem system of sanitation, water is used for removal of human excreta and other wastes. Sewage consists of two components:

  ○ The solid part, called sludge.

  ○ The liquid part, called effluent or sewage water.

Sewage is quite rich in several plant nutrients and can be used for fertilizing the crop by irrigating the soil directly with sewage water but there is a danger for the spread of several human diseases.

Chemical Fertilizers

Of the elements known to be essential for plant growth, nitrogen (N), phosphorus (P) and potassium (K) are required by plants in pretty large amounts, and are therefore, designated as major ox primary nutrients while calcium, magnesium and sulphur are secondary nutrients. For acid soils, use of Ca and Mg is necessary. Seven elements iron, manganese, boron, molybdenum, copper, zinc and chlorine are required in trace amount and hence called micro-nutrients.

Under continuous cultivation our soils are losing organic matter and mineral nutrients faster than they can be replaced. Regular loss of nutrients from the soil results in compact soil, shallow roots, increased drought, daddy and poorly productive soil. So, for the maintenance of soil fertility quick replacement of the organic matter and mineral nutrients removed from the soils is necessary.

Nitrogen (N), phosphorus (P) and potassium (K), i.e., the primary plant nutrients are commonly applied to soils in the form of commercial fertilizers and hence they are often referred to as fertilizer elements.

Table: Fertilizers and their Composition.

	Fertilizer	Composition	Source	Nitrogen%
1.	Ammonium sulphate	$(NH_4)_2SO_4$	By-produce from coke and also as synthetic	26.6
2.	Ammonium nitrate	NH_4NO_3	Synthetic	33
3.	Ammonium chloride	Dilute NH_4Cl	Synthetic	25
4.	Ammonia liquor	Dilute NH_4OH	Synthetic	20-25
5.	Ammonium phosphate	$NH_4H_2PO_4$	Synthetic	11% N and 48% P_2O_5
6.	Diammonium phosphate (DAP)	$(NH_4)_2HPO_4$	Synthetic	21% N and 53% P_2O_5

7.	Sodium nitrate	$NaNO_3$	Chile saltpetre and synthetic	16
8.	Calcium ammonium nitrate	NH_4NO_3 and Dolomite	Synthetic	20.5
9.	Calcium cyanamide	$CaCN_2$	Synthetic	21
10.	Calcium nitrate	$Ca(NO_3)_2$	Synthetic	15
11.	Urea	$CO(NH_2)_2$	Synthetic	45

Chemical fertilizers are classified into the following three group on the basis of materials supplied:

- Nitrogenous fertilizers.

- Phosphorus fertilizers.

- Potassium fertilizers.

Classification is not so simple as this grouping would imply, several such fertilizer materials as contain two of these elements, as for example, potassium nitrate and ammonium phosphate.

Nitrogenous fertilizers: Crops usually take nitrogen from the soil in the form of nitrate (NO_3^-) and ammonium ions (NH_4^+).

Nitrogen fertilizers may be divided for convenience into two groups:

- Organic nitrogen fertilizers: Organic materials such as cotton seed meal, guano and fish tank age are nitrogen carriers but because they supply less than 2% of total nitrogen added in commercial fertilizers, their use is costly and hence they are not used extensively. Nitrogen of organic fertilizers is released slowly by microbial action. They are used as special fertilizers for gardens, lawns and potted plants.

- Inorganic nitrogen fertilizers: Several inorganic chemicals are used to supply nitrogen to plants. The most important of these are presented in the table.

Phosphorus fertilizers: Phosphorus has rightly been called 'master key' to agriculture as low crop production is due more often to lack of phosphate than to the deficiency of any other element except nitrogen. In phosphorus fertilizers this element is present in the form of phosphate or superphosphate salts and it is available to the plants when it is combined with organic matter or with calcium and magnesium. Phosphorus is also found in combination with iron and aluminium and is present in certain rock minerals as apatite.

Plants take up phosphorus chiefly as phosphate $(PO_4)^-$, HPO_4^- and $H_2PO_4^-$ ions and the availability of these ions depends chiefly upon the acidity of ground. They become nearly insoluble in strongly acid or strongly alkaline soils. Release of phosphorus from phosphate rocks is slow. The breakdown of phosphate fertilizers produces phosphoric acid (P_2O_5) in soluble form that is absorbed by plants.

Use of phosphate fertilizers on alkaline soils is not suitable. Phosphorus fertilizers are classified into (i) water soluble, (ii) citrate soluble, and (iii) insoluble. When the term P.O, is used it means water soluble plus citrate soluble P_2O_5. The following are the important phosphate fertilizers being used in all parts of the world including tropical Asia.

Table: Phosphate Fertilizers.

Fertilizer with chemical composition	P_2O_5 available percentage
a) Fertilizers containing water soluble phosphorus:	
• Superphosphate Ca $(H_2PO_4)_2$ and (ordinary grade) $CaHPO_4$	16-20
• Superphosphate $Ca(H_2PO_4)_2$ and $CaHPO4$ (concentrated)	40-45
• Ammonium phosphate $NH_4H_2PO_4$	48 (11% N)
• Diammonium phosphate $(NH_4)_2HPO_4$	46-53 (4% N)
b) Citrate soluble phosphorus fertilizers:	
• Dicalcium phosphate	35-40
• Basic slag (Indian) $(CaO)_5 P_2O_5. SIO_2)$	3-5
(c) Insoluble phosphorus fertilizers:	
• Rock phosphate Fluor and chloraphatite	20-30
• Bone Meal $Ca_3(PO_4)_2$	18-20

Superphosphate

It is water soluble fertilizer. It does not affect the soil adversely. It contains mono-calcium phosphate, di-calcium phosphate and tri-calcium phosphate, gypsum, silica, iron aluminium sulphate and calcium fluoride and water.

Ammonium Phosphate

It is a fertilizer containing both nitrogen and phosphorus. It is rich in phosphoric acid content but comparatively low in nitrogen content. Ammonium superphosphate $(NH_4H_2PO_4)$. $Ca_3 (PO_4)_2$ $(NH4)_2 SO_4$. It is the cheapest fertilizer and is a mixture of nitrogen and phosphorus fertilizers. It contains nitrogen 3 to 4 per cent and phosphorus pentaoxide (P2O5) 16 to 18%.

Nitro-phosphate

It is highly hygroscopic. It contains nitrogen 13-18 per cent and phosphorus 20 per cent. It is suitable for acid soils.

Bone Meal

It is derived from bone. Bone ash and bone char are the bone products. It is suitable for acidic soils.

Basic Slag

Basic slag, a by-product in the manufacture of steel, is one of the cheapest sources of phosphorus. It is a double compound of silicate and phosphate of lime. It is dark brown in colour and alkaline in reaction.

Rock Phosphate

It occurs in natural deposits. It is light grey in colour. It is a very cheap fertilizer suitable for acid soils.

Potassium Fertilizers

These fertilizers are soluble in water which means that potassium is readily available to plants. Total potassium of potassium fertilizers is usually expressed in terms of water soluble potassium (K) or potash (K_2O). Soils of arid and semiarid areas are generally well supplied with potassium. Acid soils usually need potassium fertilizers more than neutral or alkaline soils because acid soils develop in the areas of high rainfall that leaches out available potassium. All potassium fertilizers are physiologically neutral in reaction.

The following are some common potassium fertilizers:

- Potassium chloride: It is also called muriate of potash. It contains 48—62% K_2O. It is also cheap and neutral in reaction.

- Potassium sulphate: It contains about 50% potash (K_2O). It is expensive fertilizer.

- Kainite: It is natural potassium mineral which contains 14—20% potassium. It is suitable for alkaline soils.

- Wood ashes: It is used in the form of ash as a manure. Potassium occurs in the form of potassium carbonate and the percentage of potash is from 2 to 6.5. It is suitable for alkaline soils.

Application of Micronutrients

In order to correct the deficiency of micronutrients, especially if it is very necessary, micronutrients should be added only after ascertaining the amount required. Copper, manganese, iron zinc are supplied generally as their sulphate and boron is applied as borax. Molybdenum is supplied as sodium molybdate. In recent years there has been an increase in use of chelates to supply iron, zinc, manganese and copper.

Soils vary in their ability to supply available nutrients. Some soils may be deficient in nitrogen, some may be deficient in nitrogen and phosphorus, and still some others may be deficient in nitrogen, phosphorus and potassium. To suit the variable requirements of different soils and crops, fertilizer mixtures are prepared. Fertilizer mixtures or mixed fertilizers contain two or more fertilizer materials. If the ingredients and their amounts are known, the formula is referred to as open and if they are not disclosed the formula is termed closed one.

The kinds and amounts of fertilizers to be applied to soil are determined considering the following points:

- Kinds of crop to be grown—particularly its economic value, nutrient removal and absorbing ability.

- Chemical condition of soil in respect of total nutrients and available nutrients.

- Physical state of the soil, especially its moisture content and aeration. For recommending the kind and amount of fertilizers or soil amendments, the analysis of soil is essential.

Application of Soil Conservation Practices

Loss of plant nutrients and water from the soils due to soil erosion can be checked effectively and the fertility of soil can be maintained by application of various biological and engineering methods of soil conservation. A detailed account of these methods has already been given in soil conservation topic.

Water Management Practices (Irrigation and Drainages)

Water supply is critical factor in crop production in most areas of the world. Soil moisture greatly affects the availability of mineral nutrients in the soil. It has been proved beyond doubt that fertilizer response is much higher with adequate irrigation.

Drainage and moisture control influence micronutrient availability in soils. Improving the damage of acid soils encourages the formation of less toxic oxidized forms of iron and manganese.

Prevention and Elimination of Inorganic Chemical Contamination of Soil

Loss of soil fertility due to application of toxic chemicals as pesticides can be eliminated if:

- Application of toxic chemicals to soil is reduced.
- The soil and crop are so managed as to prevent cycling of toxic chemicals.

Stabilization of Soil pH

The stabilization of pH through application of soil amendments and buffering seems to be an effective guard against the problems of non-availability of certain plant nutrients and radical changes in microbial activities arising due to change in soil pH.

Irrigation Systems

Water is a very important natural resource, which is the basis of all life forms. For sustainable agricultural production, water is one of the most precious important inputs. Plants need water in huge amount throughout their life. Water is also one of the main factors that influence most of the metabolic process such as photosynthesis, respiration, adsorption, opening and closing of stomata and translocation of food material. The growth and yield of crop plants is very much affected by the availability of water.

Water is used in Two Ways

- Withdrawal or off-channel use.
- Non-withdrawal or on-channel use.

The amount of water taken out of a streams or pumped out from underground water reservoir or surface water reservoirs to be supplied to the points of major use, such as public water supply systems, irrigation and industries, is referred to as off channel use. In non-withdrawal use, water is used without being removed from its natural source such as for navigation, swimming bodies, wildlife habitats and other recreation purposes.

Need for Irrigation

Irrigation is the artificial watering of soil to sustain plant growth. Irrigation has become necessary because of the limitation of using natural rainfall as the reliable source of water for agriculture. It is also difficult to store the rainwater for immediate irrigation purposes.

In India, monsoon is usually erratic; sometimes delayed, sometimes scarce or sometimes in excess. Through the monsoon season usually lasts for four months, most of the rains occur only during two months. Therefore, it is very difficult to plan the cropping pattern depending solely on the rams. The sufficient recharging of ground water is essential for sustainable farming, but usually this objective is not fulfilled due to unpredictable and fluctuating rains.

The efficient utilization of water resources is essential for better crop production. It includes the suitability of land and water for irrigation, planning of crops and suitable water management practices. Water management includes irrigation and drainage.

The suitable irrigation depends on:

- Time of irrigation.

- Amount of irrigation.

- Efficient method of irrigation.

Similarly, suitable drainage depends on:

- How much to drain.

- How best to drain.

- How rapidly to drain.

Stages of Crop when Irrigation is Required

The growth span of a crop plant passes through various phases and stages of growth. The rate of irrigation varies during the different stages of plant growth, i.e., from seedling to maturity stage. The growth period of irrigated crops can generally be divided into three phases, namely vegetative, flowering and maturity stage. At vegetative stage, light and intermittent irrigation is required whereas at flowering stage moderate and frequent irrigation is needed and during crop maturation stage again light irrigation is required.

Systems of Irrigation

There are different water sources from which irrigation is carried out, for example, canals, wells, open wells, tube wells, tanks etc.

Depending upon the crops, soil types, water resources, climate conditions and costs involved, several systems of irrigation are used which are as follows:

- Surface irrigation systems.

- Subsoil irrigation systems.

- Drip irrigation systems.

Surface Irrigation System

In this system, water is directly used on the surface of the soil and water follows the slope of the land, the surface irrigation system may be of following types:

- Flood irrigation: Flood irrigation is used for close-grown crops such as rice and where farm fields are leveled and water is abundant. A sheet of water is allowed to advance from different sources and remains on a field for a given period, depending on the crop, the porosity of the soil and its drainage.

Different Surface Irrigation System: (a) Flood Irrigation. (b) Furrow Irrigation. (c) Basin Irrigation.

- Basin flooding: It is used in widely spaced trees, such as in orchards, with basins built around trees and filled with water.

- Check basin: This system of irrigation is widely used in India, as it is more suitable to all types of soils and to a variety of crops.

- Furrow irrigation: In furrow irrigation, water moves in the fields in furrows, between two ridges. It is employed in the fields with row crops such as cotton and vegetables. Parallel furrows, called corrugations, are used to spread water over fields that are too irregular to flood.

Sprinkler Irrigation System

In this system, irrigation is done through pressure, to the surface of any crop or soil, in the form of a thin spray. Sprinkler irrigation system uses less water and provides better control. Each sprinkler, spaced along a pipe, sprays water in a continuous circle until the moisture reaches the root level of the crop. Centre-pivot irrigation uses long lines of sprinklers that revolve in a circular field. It is used especially for feed crops such as alfalfa, tall crops and orchards.

- Sprinkler irrigation system conveys water from the source to the field, through pipes under pressure, and distributes over the field in the form of spray of 'rain like' droplets.

- 10-20% of area is not irrigated at the comers of square of rectangular plot.

- This system requires high energy and involves huge cost of the equipment.

This system can be followed in conditions when:

- The soil is too shallow.

- The land is too steep.

- Less and frequent irrigations are needed.

- The soils are very sandy.

This system is disadvantageous in the following conditions when:

- Strong winds cause improper distribution of water.

- Evaporation losses are high from sprinkler irrigation, especially under high temperature and low relative humidity conditions.

- The initial cost is high.

Sprinkler irrigation: a. centre-pivot irrigation with long lines of sprinkler, b. sprinkler irrigation system in which water is conveyed to field through pipes under pressure.

Sub-soil Irrigation System

In this system, irrigation is done into a series of ditches in the field deeply to the impervious layer. Then it moves laterally and vertically through capillaries saturating the root/one. In this system, continuous supply of water in the root zone is assured from the artificial water table created by the ponding of irrigation water on the impervious layer. This system is very efficient because the water losses through evaporation from surface can be reduced. This system is more common in Gujarat and Jammu and Kashmir, for cash crops growing on sandy loam soils.

Drip Irrigation System

It is also called trickle irrigation. This irrigation system involves the slow application of water, drop by drop, to the root zone of a crop. It is done through mechanical devices called emitters, located at selected points along water delivery lines. Drip irrigation is extensively used in areas of acute water scarcity and especially for crops such as coconut, grape, banana, citrus, sugarcane, cotton, maize, tomato and plantation crops.

Drip Irrigation.

Advantages of Drip Irrigation

- Water loss through transpiration is low.
- It is possible to obtain better yield and quality of crops by controlling soil moisture-air-nutrient levels.
- We can save the fertilizers by monitoring the supply of nutrients as per the need of the crop.
- Improvements in biological fertility can be achieved by avoiding pollution.

Disadvantages of Conventional Irrigation Methods

- Major part of the water goes waste, only small quantity of water is utilized by the plants.
- Water is not uniformly distributed due to uneven and poor leveling of the field.
- Crops are usually subjected to cyclic changes of flooding and water stress situations by providing heavy irrigation at one time and leaving the fields to dry up for about 10 to 15 days.
- The low lying fields always get excess water that causes prolonged water logging, due to lack of leveling of fields.
- In the fields, about 10-15% of land is utilized for making channels and distributaries etc. which decreases the area of cultivation.
- Excessive irrigation and poor water management leads to water logging and accumulation of salty in the upper layer of the soil.
- In conventional irrigation, unmanageable and undesirable weeds grow.

Problems of Excess Irrigation

- Excess irrigation causes several changes in the soil and plants, resulting in reduced growth and sometimes even death of plants.

- Germinating seeds are sensitive to water logging, since they are totally dependent on the surrounding soil space for oxygen supply.

- Yield of cereals is reduced if excess irrigation is given.

- Excess water causes injury to the plant due to low oxygen supply to the root system and accumulation of toxic substances in the soil.

- Leaching of nitrates and de-nitrification occur which result in nitrogen deficiency.

- Permeability of roots decrease due to shortage of oxygen. It results in decreased uptake of water and nutrient.

SOIL RESPIRATION

Soil respiration refers to the production of carbon dioxide when soil organisms respire. This includes respiration of plant roots, the rhizosphere, microbes and fauna.

Soil respiration is a key ecosystem process that releases carbon from the soil in the form of CO_2. CO_2 is acquired from the atmosphere and converted into organic compounds in the process of photosynthesis. Plants use these organic compounds to build structural components or respire them to release energy. When plant respiration occurs below-ground in the roots, it adds to soil respiration. Over time, plant structural components are consumed by heterotrophs. This heterotrophic consumption releases CO_2 and when this CO_2 is released by below-ground organisms, it is considered soil respiration.

The amount of soil respiration that occurs in an ecosystem is controlled by several factors. The temperature, moisture, nutrient content and level of oxygen in the soil can produce extremely disparate rates of respiration. These rates of respiration can be measured in a variety of methods. Other methods can be used to separate the source components, in this case the type of photosynthetic pathway (C_3/C_4), of the respired plant structures.

Soil respiration rates can be largely affected by human activity. This is because humans have the ability to and have been changing the various controlling factors of soil respiration for numerous years. Global climate change is composed of numerous changing factors including rising atmospheric CO_2, increasing temperature and shifting precipitation patterns. All of these factors can affect the rate of global soil respiration. Increased nitrogen fertilization by humans also has the potential to affect rates over the entire planet.

Soil respiration and its rate across ecosystems is extremely important to understand. This is because soil respiration plays a large role in global carbon cycling as well as other nutrient cycles. The respiration of plant structures releases not only CO_2 but also other nutrients in those structures, such as nitrogen. Soil respiration is also associated with positive feedback with global climate change. Positive feedback is when a change in a system produces response in the same direction of the change. Therefore, soil respiration rates can be affected by climate change and then respond by enhancing climate change.

Sources of Carbon Dioxide in Soil

A portable soil respiration system measuring soil CO_2 flux.

All cellular respiration releases energy, water and CO_2 from organic compounds. Any respiration that occurs below-ground is considered soil respiration. Respiration by plant roots, bacteria, fungiand soil animals all release CO_2 in soils.

Tricarboxylic Acid (TCA) Cycle

The tricarboxylic acid (TCA) cycle – or citric acid cycle – is an important step in cellular respiration. In the TCA cycle, a six carbon sugar is oxidized. This oxidation produces the CO_2 and H_2O from the sugar. Plants, fungi, animals and bacteria all use this cycle to convert organic compounds to energy. This is how the majority of soil respiration occurs at its most basic level. Since the process relies on oxygen to occur, this is referred to as aerobic respiration.

Fermentation

Fermentation is another process in which cells gain energy from organic compounds. In this metabolic pathway, energy is derived from the carbon compound without the use of oxygen. The products of this reaction are carbon dioxide and usually either ethyl alcohol or lactic acid. Due to the lack of oxygen, this pathway is described as anaerobic respiration. This is an important source of CO_2 in soil respiration in waterlogged ecosystems where oxygen is scarce, as in peat bogs and wetlands. However, most CO_2 released from the soil occurs via respiration and one of the most important aspects of below-ground respiration occurs in the plant roots.

Root Respiration

Plants respire some of the carbon compounds which were generated by photosynthesis. When this respiration occurs in roots, it adds to soil respiration. Root respiration accounts for approximately half of all soil respiration. However, these values can range from 10–90% depending on the dominate plant types in an ecosystem and conditions under which the plants are subjected. Thus, the amount of CO_2 produced through root respiration is determined by the root biomass and specific root respiration rates. Directly next to the root is the area known as the rhizosphere, which also plays an important role in soil respiration.

Rhizosphere Respiration

The rhizosphere is a zone immediately next to the root surface with its neighboring soil. In this zone there is a close interaction between the plant and microorganisms. Roots continuously release substances, or exudates, into the soil. These exudates include sugars, amino acids, vitamins, long chain carbohydrates, enzymes and lysates which are released when roots cells break. The amount of carbon lost as exudates varies considerably between plant species. It has been demonstrated that up to 20% of carbon acquired by photosynthesis is released into the soil as root exudates. These exudates are decomposed primarily by bacteria. These bacteria will respire the carbon compounds through the TCA cycle; however, fermentation is also present. This is due to the lack of oxygen due to greater oxygen consumption by the root as compared to the bulk soil, soil at a greater distance from the root. Another important organism in the rhizosphere are root-infecting fungi or mycorrhizae. These fungi increase the surface area of the plant root and allow the root to encounter and acquire a greater amount of soil nutrients necessary for plant growth. In return for this benefit, the plant will transfer sugars to the fungi. The fungi will respire these sugars for energy thereby increasing soil respiration. Fungi, along with bacteria and soil animals, also play a large role in the decomposition of litter and soil organic matter.

Soil Animals

Soil animals graze on populations of bacteria and fungi as well as ingest and break up litter to increase soil respiration. Microfauna are made up of the smallest soil animals. These include nematodes and mites. This group specializes on soil bacteria and fungi. By ingesting these organisms, carbon that was initially in plant organic compounds and was incorporated into bacterial and fungal structures will now be respired by the soil animal. Mesofauna are soil animals from 0.1 to 2 millimeters (0.0039 to 0.0787 in) in length and will ingest soil litter. The fecal material will hold a greater amount of moisture and have a greater surface area. This will allow for new attack by microorganisms and a greater amount of soil respiration. Macrofauna are organisms from 2 to 20 millimeters (0.079 to 0.787 in), such as earthworms and termites. Most macrofauna fragment litter, thereby exposing a greater amount of area to microbial attack. Other macrofauna burrow or ingest litter, reducing soil bulk density, breaking up soil aggregates and increasing soil aeration and the infiltration of water.

Regulation of Soil Respiration

Regulation of CO_2 production in soil is due to various abiotic, or non-living, factors. Temperature, soil moisture and nitrogen all contribute to the rate of respiration in soil.

Temperature

Temperature affects almost all aspects of respiration processes. Temperature will increase respiration exponentially to a maximum, at which point respiration will decrease to zero when enzymatic activity is interrupted. Root respiration increases exponentially with temperature in its low range when the respiration rate is limited mostly by the TCA cycle. At higher temperatures the transport of sugars and the products of metabolism become the limiting factor. At temperatures over 35 °C (95 °F), root respiration begins to shut down completely. Microorganisms are divided into three temperature groups; cryophiles, mesophiles and thermophiles. Cryophiles function optimally at temperatures below 20 °C (68 °F), mesophiles function best at temperatures between 20 and 40 °C (104 °F) and thermophiles function optimally at over 40 °C (104 °F). In natural soils many different cohorts, or groups of microorganisms exist. These cohorts will all function best at different conditions, so respiration may occur over a very broad range. Temperature increases lead to greater rates of soil respiration until high values retard microbial function, this is the same pattern that is seen with soil moisture levels.

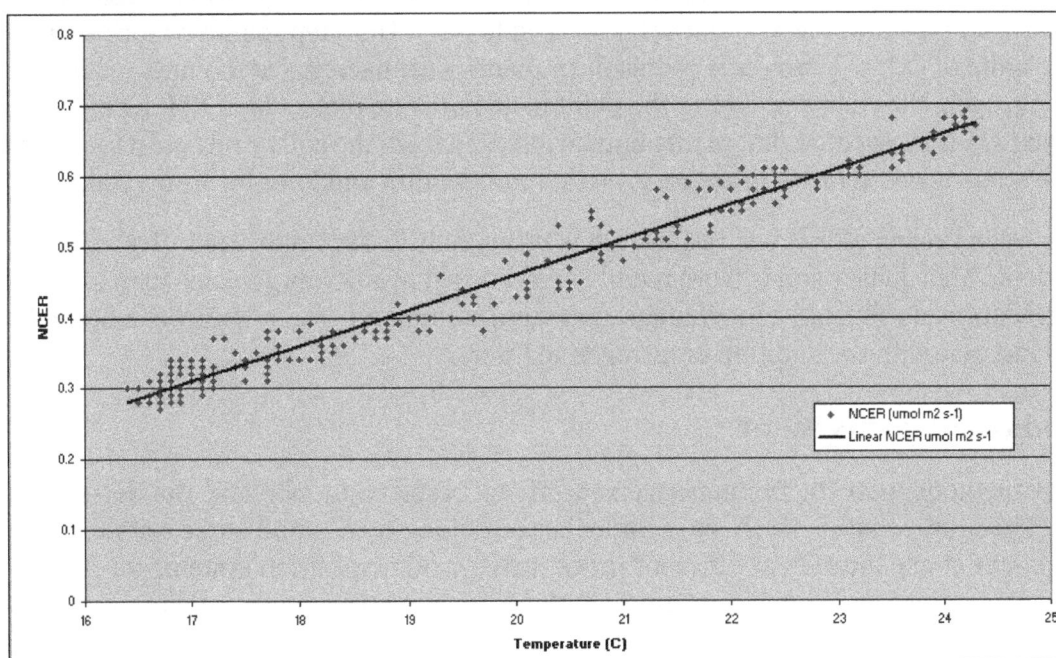

Graph showing soil respiration vs. soil temperature.

Soil Moisture

Soil moisture is another important factor influencing soil respiration. Soil respiration is low in dry conditions and increases to a maximum at intermediate moisture levels until it begins to decrease when moisture content excludes oxygen. This allows anaerobic conditions to prevail and depress aerobic microbial activity. Studies have shown that soil moisture only limits respiration at the lowest and highest conditions with a large plateau existing at intermediate soil moisture levels for most ecosystems. Many microorganisms possess strategies for growth and survival under low soil moisture conditions. Under high soil moisture conditions, many bacteria take in too much water causing their cell membrane to lyse, or break. This can decrease the rate of soil respiration temporarily, but the lysis of bacteria causes for a spike in resources for many other bacteria. This rapid increase in available labile substrates causes short-term enhanced soil respiration. Root respiration will increase with increasing soil moisture, especially in dry ecosystems; however, individual species' root respiration response to soil moisture will vary widely from species to species depending on life history traits. Upper levels of soil moisture will depress root respiration by restricting access to atmospheric oxygen. With the exception of wetland plants, which have developed specific mechanisms for root aeration, most plants are not adapted to wetland soil environments with low oxygen. The respiration dampening effect of elevated soil moisture is amplified when soil respiration also lowers soil redox through bioelectrogenesis. Soil-based microbial fuel cellsare becoming popular educational tools for science classrooms.

Nitrogen

Nitrogen directly affects soil respiration in several ways. Nitrogen must be taken in by roots in order to promote plant growth and life. Most available nitrogen is in the form of NO_3^-, which costs 0.4 units of CO_2 to enter the root because energy must be used to move it up a concentration gradient. Once inside the root the NO_3^- must be reduced to NH_3. This step requires more energy, which equals 2 units of CO_2 per molecule reduced. In plants with bacterial symbionts, which fix atmospheric nitrogen, the energetic cost to the plant to acquire one molecule of NH_3 from atmospheric N_2 is 2.36 CO_2. It is essential that plants uptake nitrogen from the soil or rely on symbionts to fix it from the atmosphere in order to assure growth, reproduction and long-term survival.

Another way nitrogen affects soil respiration is through litter decomposition. High nitrogen litter is considered high quality and is more readily decomposed by microorganisms than low quality litter. Degradation of cellulose, a tough plant structural compound, is also a nitrogen limited process and will increase with the addition of nitrogen to litter.

Methods of Measurement

Different methods exist for the measurement of soil respiration rate and the determination of sources. The most common methods include the use of long-term stand alone soil flux systems for measurement at one location at different times; survey soil respiration systems for measurement of different locations and at different times; and the use of stable isotope ratios.

Long-term Stand-alone Soil Flux Systems for Measurement at One Location over Time

An automated soil CO_2 exchange system.

These systems measure at one location over long periods of time. Since they only measure at one location, it is common to use multiple stations to reduce measuring error caused by soil variability over small distances. Soil variability may be tested with survey soil respiration instruments.

The long-term instruments are designed to expose the measuring site to ambient conditions as much as is possible between measurements.

Types of Long-term Stand-alone Instruments

Closed Mode Systems

Closed mode systems, by definition, are closed without a vent. When making measurements, closed systems take a CO_2 measurement of ambient air just after the soil sample chamber is sealed. This becomes the CO_2 reference. The system continues to take measurements at regular intervals over the next few minutes until a final CO_2 concentration is recorded at a predetermined end time over time. Closed systems have the advantages that they are used by more researchers, and they provide wind proof results. Since there is no vent, wind is not an issue. When using a closed mode system, it is important that the experimental design include a maximum measuring time for two reasons: First; if the measuring time is too long, an unacceptable positive partial CO_2 pressure can build up in the chamber limiting the amount of additional CO_2 from entering the chamber, causing a measuring artifact. The second reason is that if a chamber that is closed for too long, it can modify ambient conditions and cause measuring artifacts. Finding the optimal measuring time range is important. Tests on the soil in question can confirm the best time solution, and eliminate the partial CO_2 pressure buildup from being a significant issue.

Both individual assay information and diurnal CO_2 respiration measuring information is accessible. It is also common for such systems to also measure soil temperature, soil moisture and PAR (photosynthetically active radiation). These variables are normally recorded in the measuring file along with CO_2 values.

For determination of soil respiration and the slope of CO_2 increase, researchers have used linear regression analysis, the Pedersen (2001) algorithm, and exponential regression. There are more published references for linear regression analysis; however, the Pedersen algorithm and exponential regression analysis methods also have their following. Some systems offer a choice of mathematical methods.

When using linear regression, multiple data points are graphed and the points can be fitted with a linear regression equation, which will provide a slope. This slope can provide the rate of soil respiration with the equation $F = bV / A$, where F is the rate of soil respiration, b is the slope, V is the volume of the chamber and A is the surface area of the soil covered by the chamber. It is important that the measurement is not allowed to run over a longer period of time as the increase in CO_2 concentration in the chamber will also increase the concentration of CO_2 in the porous top layer of the soil profile. This increase in concentration will cause an underestimation of soil respiration rate due to the additional CO_2 being stored within the soil.

Open Mode Systems

Open mode systems are designed to find soil flux rates when measuring chamber equilibrium has been reached. Air flows through the chamber before the chamber is closed and sealed. This purges any non-ambient CO_2 levels from the chamber before measurement. After the chamber is closed, fresh air is pumped into the chamber at a controlled and programmable flow rate. This mixes with the CO_2 from the soil, and after a time, equilibrium is reached. The researcher specifies the equilibrium point as the difference in CO_2 measurements between successive readings, in an elapsed time. During the assay, the rate of change slowly reduces until it meets the customer's rate of change criteria, or the maximum selected time for the assay. Soil flux or rate of change is then determined once equilibrium conditions are reached within the chamber. Chamber flow rates and times are programmable, accurately measured, and used in calculations. These systems have vents that are designed to prevent a possible unacceptable buildup of partial CO_2 pressure discussed under closed mode systems. Since the air movement inside the chamber might cause increased chamber pressure, or external winds may produce reduced chamber pressure, a vent is provided that is designed to be as wind proof as possible.

Open systems are also not as sensitive to soil structure variation, or to boundary layer resistance issues at the soil surface. Air flow in the chamber at the soil surface is designed to minimize boundary layer resistance phenomena.

Hybrid Mode Systems

A hybrid system also exists. It has a vent that is designed to be as wind proof as possible, and prevent possible unacceptable partial CO_2 pressure buildup, but is designed to operate like a closed mode design system in other regards.

Survey soil respiration systems: For testing the variation of CO_2 respiration at different locations and at different times.

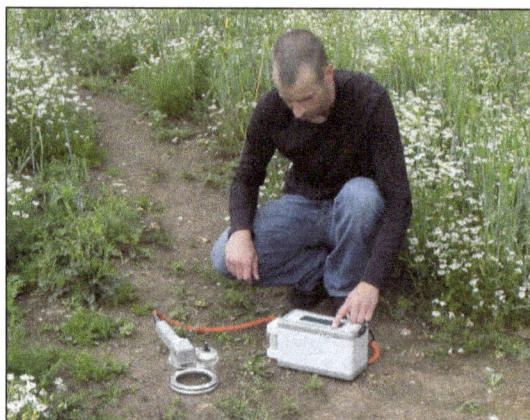

Measuring spatial variability of soil respiration in the field.

These are either open or closed mode instruments that are portable or semi-portable. They measure CO_2 soil respiration variability at different locations and at different times. With this type of instrument, soil collars that can be connected to the survey measuring instrument are inserted into the ground and the soil is allowed to stabilize for a period of time. The insertion of the soil collar temporarily disturbs the soil, creating measuring artifacts. For this reason, it is common to have several soil collars inserted at different locations. Soil collars are inserted far enough to limit lateral diffusion of CO_2. After soil stabilization, the researcher then moves from one collar to another according to experimental design to measure soil respiration.

Survey soil respiration systems can also be used to determine the number of long-term stand-alone temporal instruments that are required to achieve an acceptable level of error. Different locations may require different numbers of long-term stand-alone units due to greater or lesser soil respiration variability.

Isotope Methods

Plants acquire CO_2 and produce organic compounds with the use of one of three photosynthetic pathways. The two most prevalent pathways are the C_3 and C_4 processes. C_3 plants are best adapted to cool and wet conditions while C_4 plants do well in hot and dry ecosystems. Due to the different photosynthetic enzymes between the two pathways, different carbon isotopes are acquired preferentially. Isotopes are the same element that differ in the number of neutrons, thereby making one isotope heavier than the other. The two stable carbon isotopes are ^{12}C and ^{13}C. The C_3 pathway will discriminate against the heavier isotope more than the C_4 pathway. This will make the plant structures produced from C_4 plants more enriched in the heavier isotope and therefore root exudates and litter from these plants will also be more enriched. When the carbon in these structures is respired, the CO_2 will show a similar ratio of the two isotopes. Researchers will grow a C_4 plant on soil that was previously occupied by a C_3 plant or vice versa. By taking soil respiration measurements and analyzing the isotopic ratios of the CO_2 it can be determined whether the soil respiration is mostly old versus recently formed carbon. For example, maize, a C_4 plant, was grown on soil where spring wheat, a C_3 plant, was previously grown. The results showed respiration of C_3 SOM in the first 40 days, with a gradual linear increase in heavy isotope enrichment until day 70. The days after 70 showed a slowing enrichment to a peak at day 100. By analyzing stable carbon isotope data it is possible to determine the source components of respired SOM that was produced by different photosynthetic pathways.

Responses to Human Disturbance

Throughout the past 160 years, humans have changed land use and industrial practices, which have altered the climate and global biogeochemical cycles. These changes have affected the rate of soil respiration around the planet.

Elevated Carbon Dioxide

Since the Industrial Revolution, humans have emitted vast amounts of CO_2 into the atmosphere. These emissions have increased greatly over time and have increased global atmospheric CO_2 levels to their highest in over 750,000 years. Soil respiration increases when ecosystems are exposed to elevated levels of CO_2. Numerous free air CO_2 enrichment (FACE) studies have been conducted to test soil respiration under predicted future elevated CO_2 conditions. Recent FACE studies have shown large increases in soil respiration due to increased root biomass and microbial activity. Soil respiration has been found to increase up to 40.6% in a sweetgum forest in Tennessee and poplar forests in Wisconsin under elevated CO_2 conditions. It is extremely likely that CO_2 levels will exceed those used in these FACE experiments by the middle of this century due to increased human use of fossil fuels and land use practices.

Climate Warming

Due to the increase in temperature of the soil, CO_2 levels in our atmosphere increase, and as such the mean average temperature of the Earth is rising. This is due to human activities such as forest clearing, soil denuding, and developments that destroy autotrophic processes. With the loss of photosynthetic plants covering and cooling the surface of the soil, the infrared energy penetrates the soil heating it up and causing a rise in heterotrophic bacteria. Heterotrophs in the soil quickly degrade the organic matter and soil structure crumbles, thus it dissolves into streams and rivers into the sea. Much of the organic matter swept away in floods caused by forest clearing goes into estuaries, wetlands and eventually into the open ocean. Increased turbidity of surface waters causes biological oxygen demand and more autotrophic organisms die. Carbon dioxide levels rise with increased respiration of soil bacteria after temperatures rise due to loss of soil cover.

As mentioned earlier, temperature greatly affects the rate of soil respiration. This may have the most drastic influence in the Arctic. Large stores of carbon are locked in the frozen permafrost. With an increase in temperature, this permafrost is melting and aerobic conditions are beginning to prevail, thereby greatly increasing the rate of respiration in that ecosystem.

Changes in Precipitation

Due to the shifting patterns of temperature and changing oceanic conditions, precipitation patterns are expected to change in location, frequency and intensity. Larger and more frequent storms are expected when oceans can transfer more energy to the forming storm systems. This may have the greatest impact on xeric, or arid, ecosystems. It has been shown that soil respiration in arid ecosystems shows dynamic changes within a raining cycle. The rate of respiration in dry soil usually bursts to a very high level after rainfall and then gradually decreases as the soil dries. With an increase in rainfall frequency and intensity over area without previous extensive rainfall, a dramatic increase in soil respiration can be inferred.

Nitrogen Fertilization

Since the onset of the Green Revolution in the middle of the last century, vast amounts of nitrogen fertilizers have been produced and introduced to almost all agricultural systems. This has led to increases in plant available nitrogen in ecosystems around the world due to agricultural runoff and wind-driven fertilization. Nitrogen can have a significant positive effect on the level and rate of soil respiration. Increases in soil nitrogen have been found to increase plant dark respiration, stimulate specific rates of root respiration and increase total root biomass. This is because high nitrogen rates are associated with high plant growth rates. High plant growth rates will lead to the increased respiration and biomass found in the study. With this increase in productivity, an increase in soil activities and therefore respiration can be assured.

Importance

Soil respiration plays a significant role in the global carbon and nutrient cycles as well as being a driver for changes in climate. These roles are important to our understanding of the natural world and human preservation.

Global Carbon Cycling

Soil respiration plays a critical role in the regulation of carbon cycling at the ecosystem level and at global scales. Each year approximately 120 petagrams (Pg) of carbon are taken up by land plants and a similar amount is released to the atmosphere through ecosystem respiration. The global soils contain up to 3150 Pg of carbon, of which 450 Pg exist in wetlands and 400 Pg in permanently frozen soils. The soils contain more than four times the carbon as the atmosphere. Researchers have estimated that soil respiration accounts for 77 Pg of carbon released to the atmosphere each year. This level of release is one order of magnitude greater than the carbon release due to anthropogenic sources (6 Pg per year) such as fossil fuel burning. Thus, a small change in soil respiration can seriously alter the balance of atmosphere CO_2 concentration versus soil carbon stores. Much like soil respiration can play a significant role in the global carbon cycle, it can also regulate global nutrient cycling.

Nutrient Cycling

A major component of soil respiration is from the decomposition of litter which releases CO_2 to the environment while simultaneously immobilizing or mineralizing nutrients. During decomposition, nutrients such as nitrogen are immobilized by microbes for their own growth. As these microbes are ingested or die, nitrogen is added to the soil. Nitrogen is also mineralized from the degradation of proteins and nucleic acids in litter. This mineralized nitrogen is also added to the soil. Due to these processes, the rate of nitrogen added to the soil is coupled with rates of microbial respiration. Studies have shown that rates of soil respiration were associated with rates of microbial turnover and nitrogen mineralization. Alterations of the global cycles can further act to change the climate of the planet.

Climate Change

As stated earlier, the CO_2 released by soil respiration is a greenhouse gas that will continue to trap energy and increase the global mean temperature if concentrations continue to rise. As global temperature rises, so will the rate of soil respiration across the globe thereby leading to a higher concentration of CO_2 in the atmosphere, again leading to higher global temperatures. This is an example of a positive feedback loop. It is estimated that a rise in temperature by 2° Celsius will lead to an additional release of 10 Pg carbon per year to the atmosphere from soil respiration. This is a larger amount than current anthropogenic carbon emissions. There also exists a possibility that this increase in temperature will release carbon stored in permanently frozen soils, which are now melting. Climate models have suggested that this positive feedback between soil respiration and temperature will lead to a decrease in soil stored carbon by the middle of the 21st century.

Soil respiration is a key ecosystem process that releases carbon from the soil in the form of carbon dioxide. Carbon is stored in the soil as organic matter and is respired by plants, bacteria, fungi and animals. When this respiration occurs below ground, it is considered soil respiration. Temperature, soil moisture and nitrogen all regulate the rate of this conversion from carbon in soil organic compounds to CO_2. Many methods are used to measure soil respiration; however, the closed dynamic chamber and utilization of stable isotope ratios are two of the most prevalent techniques. Humans have altered atmospheric CO_2 levels, precipitation patterns and fertilization rates, all of which have had a significant role on soil respiration rates. The changes in these rates can alter the global carbon and nutrient cycles as well as play a significant role in climate change.

SOIL FUNCTIONS

Soils perform five key functions in the global ecosystem. Soil serves as a:

- Medium for plant growth.

- Regulator of water supplies.

- Recycler of raw materials.

- Habitat for soil organisms.

- Landscaping and engineering medium.

Soil Function - Medium for Plant Growth

As an anchor for plant roots and as a water holding tank for needed moisture, soil provides a hospitable place for a plant to take root. Some of the soil properties affecting plant growth include: Soil texture (coarse of fine), aggregate size, porosity, aeration (permeability), and water holding capacity.

Impact of soil physical properties on plant growth. The exposed roots of this corn plant show evidence of preferential growth to the right, away from where soil compaction has occurred in the wheel track area on the left.

An important function of soil is to store and supply nutrients to plants. The ability to perform this function is referred to as soil fertility. The clay and organic matter (OM) content of a soil directly influence its fertility. Greater clay and OM content will generally lead to greater soil fertility. The soil in figure has a dark brown to black color, indicating abundant OM accumulation, and a highly fertile soil.

Soil Function - Regulator of Water Supplies

As rain or snow falls upon the land, the soil is there to absorb and store the moisture for later use.

This creates a pool of available water for plants and soil organisms to live on between precipitation or irrigation events. When soils are very wet, near saturation, water moves downward through the soil profile unless it is drawn back towards the surface by evaporation and plant transpiration.

Eroded soil muddies runoff water after a spring thunderstorm in China. Soils are at risk of erosion when directly exposed to the impacts of raindrops and runoff. Surface erosion reduces the overall ability of soil to absorb and retain water.

The amount of water a soil can retain against the pull of gravity is called its water holding capacity (WHC). This property is close related to the number of very small mircro-pores present in a soil due to the effects of capillarity.

The rate of water movement into the soil (infiltration) is influenced by its texture, physical condition (soil structure and tilth), and the amount of vegetative cover on the soil surface. Coarse (sandy) soils allow rapid infiltration, but have less water storage ability, due to their generally large pore sizes. Fine textured soils have an abundance of micropores, allow them to retain a lot of water, but also causing a slow rate of water infiltration. Organic matter tends to increase the ability of all soils to retain water and also increases infiltration rates of fine textured soils.

Soil Function - Recycler of Raw Materials

As a recycler of raw materials, soil performs one of its greatest functions in the global ecosystem. Decomposition of dead plants, animals, and organisms by soil flora and fauna (e.g., bacteria, fungi, and insects) transforms their remains into simpler mineral forms, which are then utilized by other living plants, animals, and microorganisms in their creation of new living tissues and soil humus.

Many factors influence the rate of decomposition of organic materials in soil. Major determinants of the rate of decomposition include the the soil physical environment, and the chemical make-up of the decomposing materials. The activity levels of decomposing organisms are greatly impacted by the amount of water and oxygen present, and by the soil temperature. The chemical makeup of a material, especially the amount of the element nitrogen present in it, has a major impact on the 'digestibility' of any material by soil organisms. More nitrogen in the material will usually result in a faster rate of decomposition.

New life springs from remains of the old.

Plants, fungi, and insects utilize the remains of formerly living tissues to create new living tissues as a source of energy and to create new living tissues. This process recycles the dead back into the living. As dead organisms decompose, carbon dioxide gas is released and essential nutrients contained within their remains are returned to the soil to be taken up in the creation of new living tissues.

Through the processes of decomposition and humus formation, soils have the capacity to store great quantities of atmospheric carbon and essential plant nutrients. This biologically active carbon can remain in soil organic matter for decades or even centuries. This temporary storage of carbon in the organic matter of soils and biomass is termed carbon sequestration. Soil organic carbon has been identified as one of the major factors in maintaining the balance of the global carbon cycle. Land management practices that influence soil organic matter levels have been extensively studied, and are often cited as having the potential to impact the occurrence of global climate change.

Soil Function - Habitat for Soil Organisms

Soil is teeming with living organisms of varied size. Ranging from large, easily visible plant roots and animals, to very small mites and insects, to microscopically small microorganisms (e.g. bacteria and fungi.). Microorganisms are the primary decomposers of the soil, and perform much of the work of transforming and recycling old, dead materials into the raw materials needed for growth of new plants and organisms.

An earthworm in its burrow excretes its waste 'middens' on the soil surface, where
it will be further broken down by bacteria and other soil organisms. Organic materials in soil are
consumed and digested repeatedly by different organisms on their path to becoming humus.

Most living things on Earth require a few basic elements: air, food, water, and a place to live. The decomposers in soil have need of a suitable physical environment or 'habitat' to do their work. Water is necessary for the activities of all soil organisms, but they can exist in a dormant state for long periods when water is absent. Most living organisms are 'aerobic' (requiring oxygen), including plant roots and microorganisms, however some have evolved to thrive when oxygen is absent (anaerobes). Greater soil porosity and a wide range of pore sizes (diameter) in the soil allows these organisms to 'breathe' easier. Soil textural type has a great influence on the available habitat for soil organisms. Finer soils have a greater number of small 'micro-pores' that provide habitat for microorganisms like bacteria and fungi. In addition to the need for suitable habitat, all soil organisms require some type of organic material to use as an energy and carbon source, that is to say they require food. An abundant supply of fresh organic materials will ensure a robust population of soil organisms.

Landscaping and Engineering Medium

Soils are the base material for roads, homes, buildings, and other structures set upon them, but the physical properties of different soil types are greatly variable. The properties of concern in engineering and construction applications include: bearing strength, compressibility, consistency, shear strength, and shrink-swell potential. These engineering variables are influenced by the most basic soil physical properties such as texture, structure, clay mineral type, and water content. Landscaping applications range in scale from bridge and roadway construction around highway interchanges to courtyards and greenspaces around commercial sites to the grading and lawns of residential housing developments. In all these instances, both the physical and ecological functions of soils must be considered.

Exposure of soil at a construction site creates potential for soil erosion by water, wind, or both. Eroded soil pollutes waterways and causes sedimentation of ponds and reservoirs.

SOIL WATER RETENTION

Soils can process and hold considerable amount of water. They can take in water, and will keep doing so until they are full, or until the rate at which they can transmit water into and through the pores is exceeded. Some of this water will steadily drain through the soil (via gravity) and end up in the waterways and streams, but much of it will be retained, despite the influence of gravity. Much of this retained water can be used by plants and other organisms, thus contributing to land productivity and soil health.

Pores (the spaces that exist between soil particles) provide for the passage and/or retention of gasses and moisture within the soil profile. The soil's ability to retain water is strongly related to particle size; water molecules hold more tightly to the fine particles of a clay soil than to coarser particles of a sandy soil, so clays generally retain more water. Conversely, sands provide easier passage or transmission of water through the profile. Clay type, organic content, and soil structure also influence soil water retention.

The maximum amount of water that a given soil can retain is called field capacity, whereas a soil so dry that plants cannot liberate the remaining moisture from the soil particles is said to be at wilting point. Available water is that which the plants can utilize from the soil within the range between field capacity and wilting point. Roughly speaking for agriculture (top layer soil), soil is 25% water, 25% air, 45% mineral, 5% other; water varies widely from about 1% to 90% due to several retention and drainage properties of a given soil.

The role of soil water retention is profound; its effects are far reaching and relationships are invariably complex. The process by which soil absorbs water and water drains downwards is called percolation.

Soil Water Retention and Organism

Soil water retention is essential to life. It provides an ongoing supply of water to plants between periods of replenishment (infiltration), so as to allow their continued growth and survival. For example, over much of temperate Victoria, Australia, this effect is seasonal and even inter-annual; the retained soil water that has accumulated in preceding wet winters permits survival of most perennial plants over typically dry summers when monthly evaporation exceeds rainfall. Soils generally contain more nutrients, moisture and humus.

Soil Water Retention and Climate

Soil moisture has an effect on the thermal properties of a soil profile, including conductance and heat capacity. The association of soil moisture and soil thermal properties has a significant effect on temperature-related biological triggers, including seed germination, flowering, and faunal activity (More water causes soil to more slowly gain or lose temperature given equal heating; water has roughly double the Heat capacity of soil).

Recent climate modelling by Timbal et al. suggests a strong linkage between soil moisture and the persistence and variability of surface temperature and precipitation; further, that soil moisture is a significant consideration for the accuracy of "inter-annular" predications regarding the Australian climate.

Soil Water Retention, Water Balance and other Influences

The role of soil in retaining water is significant in terms of the hydrological cycle; including the relative ability of soil to hold moisture and changes in soil moisture over time:

- Soil water that is not retained or used by plants may continue downward through the profile and contribute to the water table (the permanently saturated zone at the base of the profile); this is termed "recharge". Soil that is at field capacity (among other reasons) may preclude infiltration so to increase overland flow. Both effects are associated with ground and surface water supplies, erosion, and salinity.

- Soil water can affect the structural integrity or coherence of a soil; saturated soils can become unstable and result in structural failure and mass movement. Soil water, its changes over time and management are of interest to geo-technicians and soil conservationists with an interest in maintaining soil stability.

HUMUS

In soil science, humus denominates the fraction of soil organic matter that is amorphous and without the "cellular cake structure characteristic of plants, micro-organisms or animals". Humus significantly affects the bulk density of soil and contributes to its retention of moisture and nutrients.

In agriculture, "humus" sometimes also is used to describe mature or natural compost extracted from a woodland or other spontaneous source for use as a soil conditioner. It is also used to describe a topsoilhorizon that contains organic matter (humus type, humus form, humus profile).

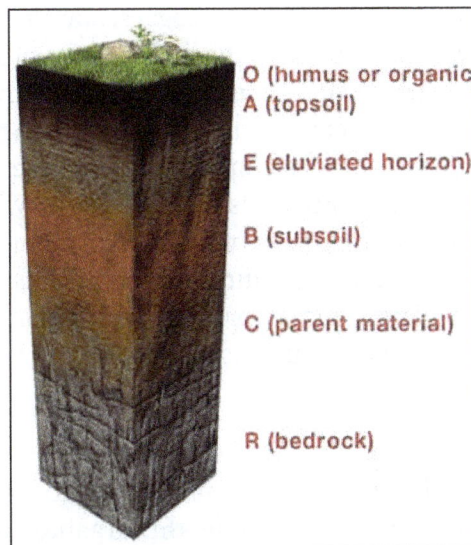

O (humus or organic
A (topsoil)

E (eluviated horizon)

B (subsoil)

C (parent material)

R (bedrock)

Humus has a characteristic black or dark brown color and is an accumulation of organic carbon. Besides the three major soil horizons of (A) surface/topsoil, (B) subsoil, and (C) substratum, some soils have an organic horizon (O) on the very surface. Hard bedrock (R) is not in a strict sense soil.

Humus is the dark organic matter that forms in soil when dead plant and animal matter decays. Humus has many nutrients that improve the health of soil, nitrogen being the most important. The ratio of carbon to nitrogen of humus is 10:1.

It is difficult to define humus precisely because it is a very complex substance which is not fully understood. Humus is different from decomposing soil organic matter. The latter looks rough and has visible remains of the original plant or animal matter. Fully humified humus, on the contrary, has a uniformly dark, spongy, and jelly-like appearance, and is amorphous; it may gradually decompose over several years or persist for millennia. It has no determinate shape, structure, or quality. However, when examined under a microscope, humus may reveal tiny plant, animal, or microbial remains that have been mechanically, but not chemically, degraded. This suggests an ambiguous boundary between humus and soil organic matter. While distinct, humus is an integral part of soil organic matter.

Humification

Microorganisms decompose a large portion of the soil organic matter into inorganic minerals that the roots of plants can absorb as nutrients. This process is termed "mineralization". In this process, nitrogen (nitrogen cycle) and the other nutrients (nutrient cycle) in the decomposed organic matter are recycled. Depending on the conditions in which the decomposition occurs, a fraction of the organic matter does not mineralize, and instead is transformed by a process called "humification" into concatenations of organic polymers. Because these organic polymers are resistant to the action of microorganisms, they are stable, and constitute *humus*. This stability implies that humus integrates into the permanent structure of the soil, thereby improving it.

Humification can occur naturally in soil or artificially in the production of compost. Organic matter is humified by a combination of saprotrophic fungi, bacteria, microbes and animals such as earthworms, nematodes, protozoa, and arthropods. Plant remains, including those that animals digested and excreted, contain organic compounds: sugars, starches, proteins, carbohydrates, lignins, waxes, resins, and organic acids. Decay in the soil begins with the decomposition of sugars and starches from carbohydrates, which decompose easily as detritivores initially invade the dead plant organs, while the remaining cellulose and lignin decompose more slowly. Simple proteins, organic acids, starches, and sugars decompose rapidly, while crude proteins, fats, waxes, and resins remain relatively unchanged for longer periods of time. Lignin, which is quickly transformed by white-rot fungi, is one of the primary precursors of humus, together with by-products of microbial and animal activity. The humus produced by humification is thus a mixture of compounds and complex biological chemicals of plant, animal, or microbial origin that has many functions and benefits in soil. Some judge earthworm humus (vermicompost) to be the optimal organic manure.

Stability

Much of the humus in most soils has persisted for more than 100 years, rather than having been decomposed into CO_2, and can be regarded as stable; this organic matter has been protected from decomposition by microbial or enzyme action because it is hidden (occluded) inside small aggregates of soil particles, or tightly sorbed or complexed to clays. Most humus that is not protected in this way is decomposed within 10 years and can be regarded as less stable or more labile. Stable humus contributes few plant-available nutrients in soil, but it helps maintain its physical structure. A very stable form of humus is formed from the slow oxidation of soil carbon after the incorporation of finely powdered charcoal into the topsoil. This process is speculated to have been important in the formation of the very fertile Amazonian *terra preta do Indio.*

Horizons

Humus has a characteristic black or dark brown color and is organic due to an accumulation of organic carbon. Soil scientists use the capital letters O, A, B, C, and E to identify the master horizons, and lowercase letters for distinctions of these horizons. Most soils have three major horizons: the surface horizon (A), the subsoil (B), and the substratum (C). Some soils have an organic horizon (O) on the surface, but this horizon can also be buried. The master horizon (E) is used for subsurface horizons that have significantly lost minerals (eluviation). Bedrock, which is not soil, uses the letter R.

Benefits of Soil Organic Matter and Humus

The importance of chemically stable humus is thought by some to be the fertility it provides to soils in both a physical and chemical sense, though some agricultural experts put a greater focus on other features of it, such as its ability to suppress disease. It helps the soil retain moisture by increasing microporosity, and encourages the formation of good soil structure. The incorporation of oxygen into large organic molecular assemblages generates many active, negatively charged sites that bind to positively charged ions (cations) of plant nutrients, making them more available to the plant by way of ion exchange. Humus allows soil organisms to feed and reproduce, and is often described as the "life-force" of the soil.

- The process that converts soil organic matter into humus feeds the population of microorganisms and other creatures in the soil, and thus maintains high and healthy levels of soil life.

- The rate at which soil organic matter is converted into humus promotes (when fast) or limits (when slow) the coexistence of plants, animals, and microorganisms in the soil.

- Effective humus and stable humus are additional sources of nutrients for microbes: the former provides a readily available supply and the latter acts as a longeval storage reservoir.

- Decomposition of dead plant material causes complex organic compounds to be slowly oxidized (lignin-like humus) or to decompose into simpler forms (sugars and amino sugars, and aliphatic and phenolic organic acids), which are further transformed into microbial biomass (microbial humus) or reorganized, and further oxidized, into humic assemblages (fulvic acids and humic acids), which bind to clay minerals and metal hydroxides. The ability of plants to absorb humic substances with their roots and metabolize them has been long debated. There is now a consensus that humus functions hormonally rather than simply nutritionally in plant physiology.

- Humus is a colloidal substance and increases the cation exchange capacity of soil, hence its ability to store nutrients by chelation. While these nutrient cations are available to plants, they are held in the soil and prevented from being leached by rain or irrigation.

- Humus can hold the equivalent of 80–90% of its weight in moisture, and therefore increases the soil's capacity to withstand drought.

- The biochemical structure of humus enables it to moderate, i. e. buffer, excessive acidic or alkaline soil conditions.

- During humification, microbes secrete sticky, gum-like mucilages; these contribute to the crumby structure (tilth) of the soil by adhering particles together and allowing greater aeration of the soil. Toxic substances such as heavy metals and excess nutrients can be chelated, i. e., bound to the organic molecules of humus, and so prevented from leaching away.

- The dark, usually brown or black, color of humus helps to warm cold soils in Spring.

SOIL HORIZON

A cross section of a soil, revealing horizons.

In most soil classification systems, horizons are used to define soil types. The German system uses entire horizon sequences for definition. Other systems pick out certain horizons, the "diagnostic horizons", for the definition; examples are the World Reference Base for Soil Resources (WRB), the USDA soil taxonomyand the Australian Soil Classification. Diagnostic horizons are usually indicated with names, e.g. the "cambic horizon" or the "spodic horizon". The WRB lists 37 diagnostic horizons. In addition to these diagnostic horizons, some other soil characteristics may be needed to define a soil type. Some soils do not have a clear development of horizons.

A soil horizon *sensu stricto* is a result of soil-forming processes (pedogenesis). Layers that do not have undergone such processes may be simply called "layers". Some soil scientists use the word layer in a more general way, including the horizons *sensu stricto*.

Horizon Sequence

Many soils have an organic surface layer, which is denominated with a capital letter (different letters, depending from the system). The mineral soil usually starts with an A horizon. If a well-developed subsoil horizon as a result of soil formation exists, it is generally called a B horizon. An underlying loose, but poorly developed horizon is called a C horizon. Hard bedrock is mostly denominated R. Most individual systems defined more horizons and layers than just these five. In the following, the horizons and layers are listed more or less by their position from top to bottom within the soil profile. Not all of them are present in every soil.

Soils with a history of human interference, for instance through major earthworks or regular deep ploughing, may lack distinct horizons almost completely. When examining soils in the field, attention must be paid to the local geomorphology and the historical uses, to which the land has been put, in order to ensure that the appropriate names are applied to the observed horizons.

Example of a Soil Profile

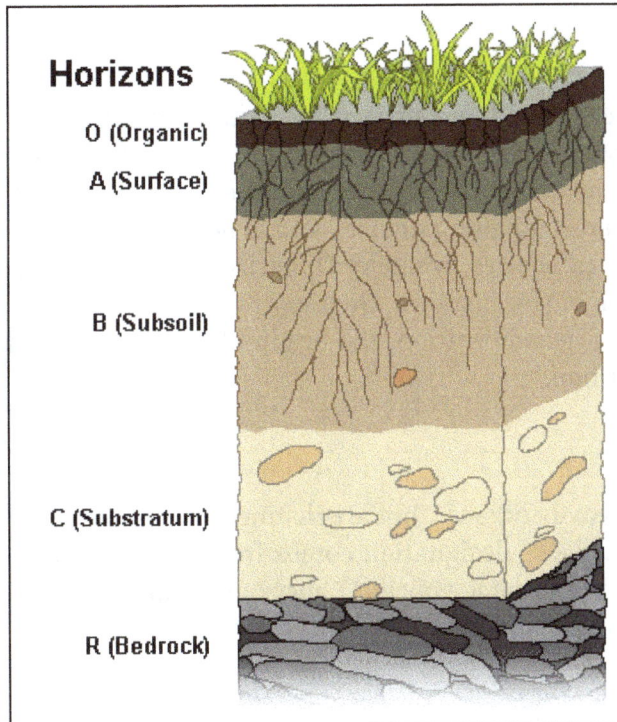

- Organic surface layer: Litter layer of plant residues, the upper part often relatively undecomposed, but the lower part may be strongly humified.

- Surface soil: Layer of mineral soil with most organic matter accumulation and soil life. Additionally, due to weathering, oxides (mainly iron oxides) and clay minerals are formed and accumulated. It has a pronounced soil structure. But in some soils, clay minerals, iron, aluminium, organic compounds, and other constituents are soluble and move downwards. When this eluviation is pronounced, a lighter coloured E subsurface soil horizon is apparent at the base of the A horizon. A horizons may also be the result of a combination of soil bioturbation and surface processes that winnow fine particles from biologically mounded topsoil. In this case, the A horizon is regarded as a "biomantle".

- Subsoil: This layer has normally less organic matter than the A horizon, so its colour is mainly derived from iron oxides. Iron oxides and clay minerals accumulate as a result of weathering. In a soil, where substances move down from the topsoil, this is the layer, where they accumulate. The process of accumulation of clay minerals, iron, aluminium and organic compounds, is referred to as illuviation. The B horizon has generally a soil structure.

- Substratum: Layer of non-indurated poorly weathered or unweathered rocks. This layer may accumulate the more soluble compounds like $CaCO_3$. Soils formed in situ from non-indurated material exhibit similarities to this C layer.

- Bedrock: R horizons denote the layer of partially weathered or unweathered bedrock at the base of the soil profile. Unlike the above layers, R horizons largely comprise continuous masses (as opposed to boulders) of hard rock that cannot be excavated by hand. Soils formed in situ from bedrock will exhibit strong similarities to this bedrock layer.

Horizons

O Horizon

The "O" stands for organic matter. It is a surface layer, dominated by the presence of large amounts of organic matter in varying stages of decomposition. In the Australian system, the O horizon should be considered distinct from the layer of leaf litter covering many heavily vegetated areas, which contains no weathered mineral particles and is not part of the soil itself. O horizons may be divided into O_1 and O_2 categories, whereby O_1 horizons contain undecomposed matter whose origin can be spotted on sight (for instance, fragments of leaves), and O_2 horizons contain organic debris in various stages of decomposition, the origin of which is not readily visible. O horizons contain ≥ 20% organic carbon.

P Horizon

These horizons are also heavily organic, but are distinct from O horizons in that they form under waterlogged conditions. The "P" designation comes from their common name, peats. They may be divided into P_1 and P_2 in the same way as O horizons. P horizons contain ≥ 12 to 18% organic carbon, depending on the clay content.

A Horizon

Surface runoff, a potential pathway for nonpoint
source pollution, from a farm field in Iowa during a rain storm.

The A horizon is the top layer of the mineral soil horizons, often referred to as 'topsoil'. This layer contains dark decomposed organic matter, which is called "humus". The technical definition of an

A horizon may vary between the systems, but it is most commonly described in terms relative to deeper layers. "A" horizons may be darker in colour than deeper layers and contain more organic matter, or they may be lighter but contain less clay or pedogenic oxides. The A is a surface horizon, and as such is also known as the zone in which most biological activity occurs. Soil organisms such as earthworms, potworms (enchytraeids), arthropods, nematodes, fungi, and many species of bacteria and archaea are concentrated here, often in close association with plant roots. Thus, the A horizon may be referred to as the biomantle. However, since biological activity extends far deeper into the soil, it cannot be used as a chief distinguishing feature of an A horizon. The A horizon may be further subdivided into A1 (dark, maximum biologic activity), A2 (paler) and A3 (transitional to the B horizon).

E Horizon

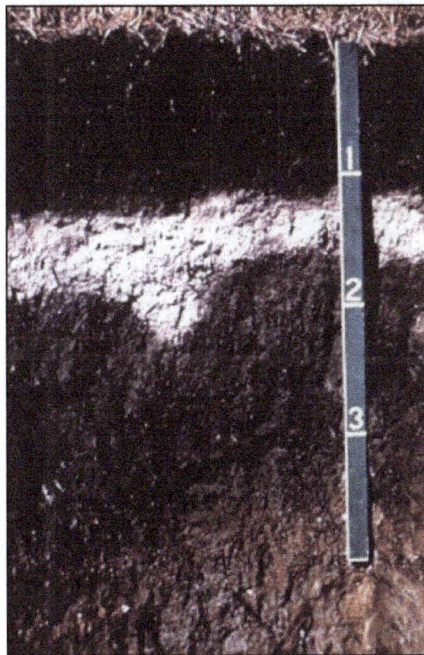

Albic Luvisol – dark surface horizon on a bleached subsurface horizon
(an albic horizon) that tongues into a clay illuviation (Bt) horizon.

"E" (not used in the Australian System) being short for eluviated, is most commonly used to label a horizon that has been significantly leached of its mineral and/or organic content, leaving a pale layer largely composed of silicates or silica. These are present only in older, well-developed soils, and generally occur between the A and B horizons. In systems where (like in the Australian system) this designation is not employed, leached layers are classified firstly as an A or B according to other characteristics, and then appended with the designation "e". In soils that contain gravels, due to animal bioturbation, a stone layer commonly forms near or at the base of the E horizon.

B Horizon

The B horizon is commonly referred to as "subsoil" and consists of mineral layers which are significantly altered by pedogenesis, mostly with the formation of iron oxides and clay minerals. It is usually brownish or reddish due to the iron oxides, which increases the chroma of the subsoil to a degree that it can be distinguished from the other horizons. The weathering may be biologically mediated. In addition, the B horizon is defined as having a distinctly different structure or consistency than the horizons above and the horizons below.

The B horizon can also accumulate minerals and organic matter that are migrating downwards from the A and E horizons. If so, this layer is also known as the illuviated or illuvial horizon.

As with the A horizon, the B horizon may be divided into B1, B2, and B3 types under the Australian system. B1 is a transitional horizon of the opposite nature to an A3 – dominated by the properties of the B horizons below it, but containing some A-horizon characteristics. B2 horizons have a high concentration of clay minerals or oxides. B3 horizons are transitional between the overlying B layers and the material beneath it, whether C or D horizon.

The A3, B1, and B3 horizons are not tightly defined, and their use is generally at the discretion of the individual worker. Plant roots penetrate throughout this layer, but it has very little humus.

The A/E/B horizons are referred to collectively as the "solum", the surface depth of the soil where biologically activity and climate effects drives pedogenesis. The layers below the solum have no collective name but are distinct in that they are noticeably less affected by surface soil-forming processes.

C Horizon

The C horizon is below the solum horizons. This layer is little affected by pedogenesis. Clay illuviation, if present, is not significant. The absence of solum-type development (pedogenesis) is one of the defining attributes. The C horizon forms either in deposits (e.g., loess, flood deposits, landslides) or it formed from weathering of residual bedrock. The C horizon may be enriched with carbonates carried below the solum by leaching. If there is no lithologic discontinuity between the solum and the C horizon and no underlying bedrock present, the C horizon resembles the parent material of the solum.

Soil with broken rock fragments overlying bedrock, Sandside Bay, Caithness.

D Horizon

D horizons are not universally distinguished, but in the Australian system refer to "any soil material below the solum that is unlike the solum in its general character, is not C horizon, and cannot be given reliable horizon designation. It may be recognized by the contrast in pedologic organization between it and the overlying horizons".

R Horizon

R horizons denote the layer of partially weathered or unweathered bedrock at the base of the soil profile. Unlike the above layers, R horizons largely comprise continuous masses (as opposed to boulders) of hard rock that cannot be excavated by hand. If there is no lithologic discontinuity between the solum and the R horizon, the R horizon resembles the parent material of the solum.

L Horizon

L (Limnic) horizons or layers indicate mineral or organic material that has been deposited in water by precipitation or through the actions of aquatic organisms. Included are coprogenous earth (sedimentary peat), diatomaceous earth, and marl; and is usually found as a remnant of past bodies of standing water.

Transitional Horizons

A horizon that combines the characteristics of two horizons is indicated with both capital letters, the dominant one written first. Example: AB and BA. If distinct parts have properties of two kinds of horizons, the horizon symbols are combined using a slash (/). Example: A/B and B/A.

Horizon Suffixes

In addition to the main descriptors above, several modifiers exist to add necessary detail to each horizon. Firstly, each major horizon may be divided into sub-horizons by the addition of a numerical subscript, based on minor shifts in colour or texture with increasing depth (e.g., B21, B22, B23 etc.). While this can add necessary depth to a field description, workers should bear in mind that excessive division of a soil profile into narrow sub-horizons should be avoided. Walking as little as ten metres in any direction and digging another hole can often reveal a very different profile in regards to the depth and thickness of each horizon. Over-precise description can be a waste of time. In the Australian system, as a rule of thumb, layers thinner than 5 cm (2 inches) or so are best described as pans or segregations within a horizon rather than as a distinct layer.

Suffixes describing particular features of a horizon may also be added. The Australian system provides the following suffixes:

- B: Buried horizon.
- C: Presence of mineral concretions or nodules, perhaps of iron, aluminium, or manganese.
- D: Root restricting layer.
- E: Conspicuously bleached.
- F: Faunal accumulations in A horizons.
- G: Gleyed horizon.

- H: Accumulation of organic matter.

- J: Sporadically bleached.

- K: Accumulation of carbonates, commonly calcium carbonate.

- M: Strong cementation or induration.

- P: Disturbed by ploughing or other tillage practices (A horizons only).

- Q: Accumulation of secondary silica.

- R: Weathered, digable rock.

- S: Sesquioxide accumulation.

- T: Accumulation of clay minerals.

- W: Weak development.

- X: Fragipan.

- Y: Accumulation of calcium sulfate (gypsum).

- Z: Accumulation of salts more soluble than calcium sulfate.

Buried Soils

Soil formation is often described as occurring *in situ*: Rock breaks down, weathers and is mixed with other materials, or loose sediments are transformed by weathering. But the process is often far more complicated. For instance, a fully formed profile may have developed in an area only to be buried by wind- or water-deposited sediments which later formed into another soil profile. This sort of occurrence is most common in coastal areas, and descriptions are modified by numerical prefixes. Thus, a profile containing a buried sequence could be structured O, A1, A2, B2, 2A2, 2B21, 2B22, 2C with the buried profile commencing at 2A2.

Master Horizons and Layers

- O: Organic soil materials (not limnic).

- A: Mineral; organic matter (humus) accumulation.

- E: Mineral; some loss of Fe, Al, clay, or organic matter.

- B: Subsurface accumulation of clay, Fe, Al, Si, humus, $CaCO_3$, $CaSO_4$; or loss of $CaCO_3$; or accumulation of sesquioxides; or subsurface soil structure.

- C: Little or no pedogenic alteration, unconsolidated earthy material, soft bedrock.

- L: Limnic soil materials.

- W: A layer of liquid water (W) or permanently frozen water (Wf) within or beneath the soil (excludes water/ice above soil).

- M: Root-limiting subsoil layers of human-manufactured materials.
- R: Bedrock, strongly cemented to indurated.

Transitional Horizons and Layers

A horizon that combines the characteristics of two master horizons is indicated with both capital letters, the dominant one written first. Example: AB and BA. If discrete, intermingled bodies of two master horizons occur together, the horizon symbols are combined using a slash (/). Example: A/B and B/A.

Horizon Suffixes

- a: Highly decomposed organic matter (used only with O).
- aa: (proposed) Accumulation of anhydrite ($CaSO_4$).
- b: Buried genetic horizon (not used with C horizons).
- c: Concretions or nodules.
- co: Coprogenous earth (used only with L).
- d: Densic layer (physically root restrictive).
- di: Diatomaceous earth (used only with L).
- e: Moderately decomposed organic matter (used only with O).
- f: Permanently frozen soil or ice (permafrost); continuous subsurface ice; not seasonal ice.
- ff: Permanently frozen soil ("dry" permafrost); no continuous ice; not seasonal ice.
- g: Strong gley.
- h: Illuvial organic matter accumulation.
- i: Slightly decomposed organic matter (used only with O).
- j: Jarosite accumulation.
- jj: Evidence of cryoturbation.
- k: Pedogenic $CaCO_3$ accumulation (<50% by vol.).
- kk: Major pedogenic $CaCO_3$ accumulation (≥50% by vol.).
- m: Continuous cementation (pedogenic).
- ma: Marl (used only with L).
- n: Pedogenic, exchangeable sodium accumulation.
- o: Residual sesquioxide accumulation (pedogenic).
- p: Plow layer or other artificial disturbance.
- q: Secondary (pedogenic) silica accumulation.
- r: Weathered or soft bedrock.

- s: Illuvial sesquioxide and organic matter accumulation.

- se: Presence of sulfides (in mineral or organic horizons).

- ss: Slickensides.

- t: Illuvial accumulation of silicate clay.

- u: Presence of human-manufactured materials (artifacts).

- v: Plinthite.

- w: Weak color or structure within B (used only with B).

- x: Fragipan characteristics.

- y: Accumulation of gypsum.

- yy: Dominance of gypsum ($\approx \geq 50\%$ by vol.).

- z: Pedogenic accumulation of salt more soluble than gypsum.

Other Horizon Modifiers

Numerical prefixes are used to denote lithologic discontinuities. By convention, 1 is not shown. Numerical suffixes are used to denote subdivisions within a master horizon. Example: A, E, Bt1, 2Bt2, 2BC, 3C1, 3C2.

Horizons and Layers according to the FAO Guidelines for Soil Description

The World Reference Base for Soil Resources (WRB) recommends the use of these horizon denominations.

Master Horizons and Layers

- H Horizons or Layers: These are layers of organic material. Organic material is defined by having a certain minimum content of soil organic carbon. In the WRB, this is 20% (by weight). The H horizon is formed from organic residues that are not incorporated into the mineral soil. The residues may be partially altered by decomposition. Contrary to the O horizons, the H horizons are saturated with water for prolonged periods, or were once saturated but are now drained artificially. In many H horizons, the residues are predominantly mosses. Although these horizons form above the mineral soil surface, they may be buried by mineral soil and therefore be found at greater depth. H horizons may be overlain by O horizons that especially form after drainage.

- O horizons or Layers: These are layers of organic material. Organic material is defined by having a certain minimum content of soil organic carbon. In the WRB, this is 20% (by weight). The O horizon is formed from organic residues that are not incorporated into the mineral soil. The residues may be partially altered by decomposition. Contrary to the H horizons, the O horizons are not saturated with water for prolonged periods and not drained artificially. In many O horizons, the residues are leaves, needles, twigs, moss, and lichens. Although these horizons form above the mineral soil surface, they may be buried by mineral soil and therefore be found at greater depth.

- A Horizons: These are mineral horizons that formed at the surface or below an O horizon. All or much of the original rock structure has been obliterated. Additionally, they are characterized by one or more of the following:

  ○ An accumulation of humified organic matter, intimately mixed with the mineral fraction, and not displaying properties characteristic of E or B horizons;

  ○ Properties resulting from cultivation, pasturing, or similar kinds of disturbance;

  ○ A morphology that is different from the underlying B or C horizon, resulting from processes related to the surface.

If a surface horizon has properties of both A and E horizons but the dominant feature is an accumulation of humified organic matter, it is designated an A horizon.

- E Horizons: These are mineral horizons in which the main feature is loss of clay minerals, iron, aluminium, organic matter or some combination of these, leaving a concentration of sand and silt particles. However, pedogenesis is advanced, because the lost substances first have been formed or accumulated there. All or much of the original rock structure is obliterated. An E horizon is usually, but not necessarily, lighter in colour than an underlying B horizon. In some soils, the colour is that of the sand and silt particles. An E horizon is most commonly differentiated from an underlying B horizon: by colour of higher value or lower chroma, or both; by coarser texture; or by a combination of these properties. An E horizon is commonly near to the surface, below an O or A horizon and above a B horizon. However, the symbol E may be used without regard to the position in the profile for any horizon that meets the requirements and that has resulted from soil genesis.

- B Horizons: These are horizons that formed below an A, E, H or O horizon, and in which the dominant features are the obliteration of all or much of the original rock structure, together with one or a combination of the following:

  ○ Residual concentration of oxides (especially iron oxides) and/or clay minerals;

  ○ Evidence of removal of carbonates or gypsum;

  ○ Illuvial concentration, alone or in combination, of clay minerals, iron, aluminium, organic matter, carbonates, gypsum or silica;

  ○ Coatings of oxides that make the horizon conspicuously lower in value, higher in chroma, or redder in hue than overlying and underlying horizons without apparent illuviation of iron;

  ○ Alteration that forms clay minerals or liberates oxides or both and that forms a granular, blocky or prismatic structure if volume changes accompany changes in moisture content;

  ○ Brittleness.

All kinds of B horizons are, or were originally, subsurface horizons.

Examples of layers that are not B horizons are: layers in which clay films either coat rock fragments or are found on finely stratified unconsolidated sediments, whether the films were formed in place or by illuviation; layers into which carbonates have been illuviated but that are not contiguous to an overlying genetic horizon; and layers with gleying but no other pedogenic changes.

- C horizons or layers: These are horizons or layers, excluding hard bedrock, that are little affected by pedogenic processes and lack properties of H, O, A, E or B horizons. Most are mineral layers, but some siliceous and calcareous layers, such as shells, coral and diatomaceous earth, are included. The material of C layers may be either like or unlike that from which the overlying solum presumably formed. Plant roots can penetrate C horizons, which provide an important growing medium. Included as C layers are sediments, saprolite, non-indurated bedrock and other geological materials that commonly slake within 24 hours, when air-dry or drier chunks are placed in water, and that, when moist, can be dug with a spade. Some soils form in material that is already highly weathered, and if such material does not meet the requirements of A, E or B horizons, it is designated C. Changes not considered pedogenic are those not related to overlying horizons. Layers having accumulations of silica, carbonates or gypsum, even if indurated, may be included in C horizons, unless the layer is obviously affected by pedogenic processes; then it is a B horizon.

- R layers: These consist of hard bedrock underlying the soil. Granite, basalt, quartzite and indurated limestone or sandstone are examples of bedrock that are designated R. Air-dry or drier chunks of an R layer, when placed in water, will not slake within 24 hours. The R layer is sufficiently coherent when moist to make hand digging with a spade impractical. The bedrock may contain cracks, but these are so few and so small that few roots can penetrate. The cracks may be coated or filled with soil material.

- I layers: These are ice lenses and wedges that contain at least 75 percent ice (by volume) and that distinctly separate layers (organic or mineral) in the soil.

- L layers: These are sediments deposited in a body of water. They may be organic or mineral. Limnic material is either: (i) Deposited by precipitation or through action of aquatic organisms, such as algae, especially diatoms; or (ii) Derived from underwater and floating aquatic plants and subsequently modified by aquatic animals. L layers include coprogenous earth or sedimentary peat (mostly organic), diatomaceous earth (mostly siliceous), and marl (mostly calcareous).

- W layers: These are either water layers in soils or water layers submerging soils. The water is present either permanently or cyclic within the time frame of 24 hours. Some organic soils float on water. In other cases, shallow water (i.e. water not deeper than 1 m) may cover the soil permanently, as in the case of shallow lakes, or cyclic, as in tidal flats. The occurrence of tidal water can be indicated by the letter W in brackets: (W).

Transitional Horizons and Layers

A horizon that combines the characteristics of two master horizons is indicated with both capital letters, the dominant one written first. Example: AB and BA. If discrete, intermingled bodies of two master horizons occur together, the horizon symbols are combined using a slash (/). Example: A/B and B/A. The master horizon symbols may be followed by the lowercase letters indicating subordinate characteristics. Example: AhBw. The I, L and W symbols are not used in transitional horizon designations.

Subordinate Characteristics

This is the list of suffixes to the master horizons. After the hyphen, it is indicated to which master horizons the suffixes can be added.

- a: Highly decomposed organic material - H and O horizons.

- b: Buried genetic horizon - mineral horizons, not cryoturbated.

- c: Concretions or nodules - mineral horizons.

- c: Coprogenous earth - L horizon.

- d: Dense layer (physically root restrictive) - mineral horizons, not with m.

- d: Diatomaceous earth - L horizon.

- e: Moderately decomposed organic material - H and O horizons.

- f: Frozen soil - not in I and R horizons.

- g: Stagnic conditions - no restriction.

- h: Accumulation of organic matter - mineral horizons.

- i: Slickensides - mineral horizons.

- i: Slightly decomposed organic material - H and O horizons.

- j: Jarosite accumulation - no restriction.

- k: Accumulation of pedogenic carbonates - no restriction.

- l: Mottling due to upmoving groundwater (gleying) - no restriction.

- m: Strong cementation or induration (pedogenic, massive) - mineral horizons.

- m: Marl - L horizon.

- n: Pedogenic accumulation of exchangeable sodium - no restriction.

- o: Residual accumulation of sesquioxides (pedogenic) - no restriction.

- p: Ploughing or other human disturbance - no restriction; ploughed E, B, or C horizons are referred to as Ap.

- q: Accumulation of pedogenic silica - no restriction.

- r: Strong reduction - no restriction.

- s: Illuvial accumulation of sesquioxides - B horizons.

- t: Illuvial accumulation of clay minerals - B and C horizons.

- u: Urban and other human-made materials artefacts - H, O, A, E, B and C horizons.

- v: Occurrence of plinthite - no restriction.

- w: Development of colour or structure - B horizons.

- x: Fragipan characteristics - no restriction.

- y: Pedogenic accumulation of gypsum - no restriction.

- z: Pedogenic accumulation of salts more soluble than gypsum - no restriction.

- @: Evidence of cryoturbation - no restriction.

Discontinuities and Vertical Subdivisions

Numerical prefixes are used to denote lithic discontinuities. By convention, 1 is not shown. Numerical suffixes are used to denote subdivisions within a horizon. The horizons in a profile are combined using a hyphen (-). Example: Ah-E-Bt1-2Bt2-2BwC-3C1-3C2.

Diagnostic Soil Horizons

Many soil classification systems have diagnostic horizons. A diagnostic horizon is a horizon used to define soil taxonomic units (e.g. to define soil types). The presence or absence of one or more diagnostic horizons in a required depth is used for the definition of a taxonomic unit. In addition, most classification systems use some other soil characteristics for the definition of taxonomic units. The diagnostic horizons need to be thoroughly defined by a set of criteria. When allocating a soil (a pedon, a soil profile) to a taxonomic unit, one has to check every horizon of this soil and decide, whether or not the horizon fulfils the criteria of a diagnostic horizon. Based on the identified diagnostic horizons, one can proceed with the allocation of the soil to a taxonomic unit. In the following, the diagnostic horizons of two soil classification systems are listed.

References

- Soil-texture-definition-classes-and-its-determination: soilmanagementindia.com, Retrieved 5 August, 2019

- Soil Science Glossary Terms Committee (2008). Glossary of Soil Science Terms 2008. Madison, WI: Soil Science Society of America. ISBN 978-0-89118-851-3

- Soil-colour-factors-determination-and-its-implication, soil-colour: soilmanagementindia.com, Retrieved 6 January, 2019

- Samouëlian, A; Cousin, I; Richard, G; Tabbagh, A; Bruand, A. (2003). "Electrical resistivity imaging for detecting soil cracking at the centimetric scale". Soil Science Society of America Journal. 67 (5): 1319–1326. Bibcode:2003SSASJ..67.1319S. Doi:10.2136/sssaj2003.1319. Archived from the original on 2010-06-15

- Soil-structure-definition-types-and-formation, soil: soilmanagementindia.com, Retrieved 7 February, 2019

- Chesworth, Ward (2008). Encyclopedia of soil science. Dordrecht, Netherlands: Springer. P. 694. ISBN 978-1402039942. Retrieved 2 July 2016

- Allison, Steven D.; Wallenstein, Matthew D.; Bradford, Mark A. (2010). "Soil-carbon response to warming dependent on microbial physiology". Nature Geoscience. 3 (5): 336–340. Bibcode:2010natge...3..336A. Doi:10.1038/ngeo846

- Soil-resistivity: circuitglobe.com, Retrieved 8 March, 2019

Branches of Soil Science

There are various branches of soil science which deals with the constitution, properties and processes taking place in soil. Some of them are soil physics, soil chemistry, soil biology, soil microbiology, soil ecology, pedology, edaphology, etc. All these diverse branches of soil science have been carefully analyzed in this chapter.

SOIL PHYSICS

Soil physics addresses physical properties and processes of soil and porous medium systems. Properties such as texture, structure, density, surface area, and aggregate stability are addressed along with water-content and potential leading to the water retention character of soil. Processes involving transport of heat, solutes, gasses and of course water are characterized and modeled.

There are many different themes and interactions that soil physicists study, some involving the transfer of energy from one place to another, others study how water moves and interacts with plants, and others study gas emissions, others find out what happens when soil is tilled, and what happens if the soils become to salty to use.

Water Movement

Measuring saturated hydraulic conductivity This is a device (Amoozemeter) that measures how fast water flows in the ground.

Since water is vitally important for plant growth and drinking, it is important to understand how it moves through the soil matrix (the gigantic maze of aggregates and soil particles). Understanding how water moves can help make decisions about land use and which crops to grow. If a soil has a lot of sand, water moves through it very quickly. This may mean that fertilizers, waste products, or chemicals move through the soil into the groundwater very quickly. If soils have too much clay, the water may not infiltrate into the soils fast enough, and waste products or fertilizers can end up running off into streams and lakes. Some soils can hold onto water well and store it below ground so it is available for plant roots.

Heat Movement and Evaporation

Soil temperature is really important for biological activity and soil formation. Temperatures that are really cold or too hot have limited plant growth and microbial activity, and less chemical action. A dark colored soil absorbs more solar radiation compared to a lighter soil. This changes how plants and microbes can live. How soils conduct heat determines how water evaporates from soil, how much radiation bounces back into the soil atmosphere, how the air is heated above the soil, and how hot the soil is itself. This rate is closely linked with the soil's water content. Hot soils create a lot of microbial decomposition, leading to less organic matter, and they also demand that plants transpire (breathe) faster, which is not a very good thing if the climate is dry, or if there is a drought.

Gas Emissions and Air Movement

Soils are a dynamic system of pores and texture, with spaces inbetween that air and water flow through. Soils have the ability to store greenhouse gases, like carbon dioxide, methane, and nitrous oxide, but they can also release them. Whether soils are a sink (storing greenhouse gases) or a source (releasing greenhouse gases) depends on how the soil is managed. Soil scientists who study gas emission and air movement from soils try to increase the organic matter content in soils by using conservation tillage practices, reducing erosion, and comparing different cropping systems.

SOIL CHEMISTRY

Soil chemistry is the study of the chemical characteristics of soil. Soil chemistry is affected by mineral composition, organic matter and environmental factors. The topics studied under soil chemistry are:

Ion Exchanges

Ion exchange involves the movement of cations (positively charged elements like calcium, magnesium, and sodium) and anions (negatively charged elements like chloride, and compounds like nitrate) through the soils. In the United States, cation exchange is much more common.

Cation exchange is the interchanging between a cation in the solution of water around the soil particle, and another cation that is stuck to the clay surface. The number of cations in the soil water solution is much smaller than the number that is attached to soil particles.

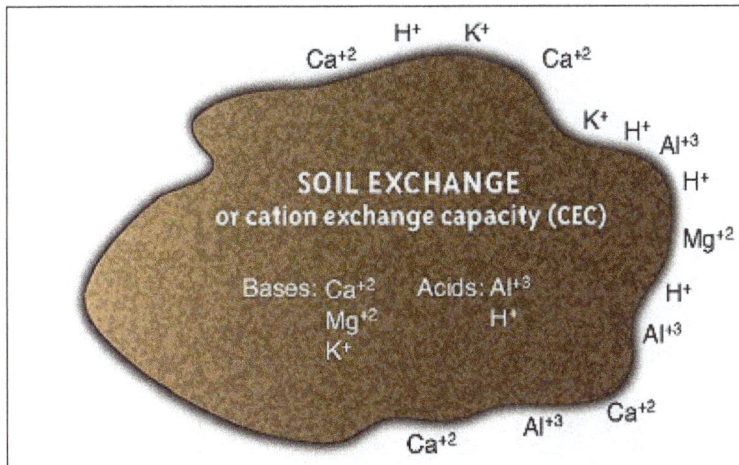

This diagram represents soil cations attached firmly to the soil.

The total amount of positive charges that the soil can absorb is called the cation exchange capacity (CEC). CEC impacts how quickly nutrients move through the profile. A soil with a low CEC is much less fertile because it cannot hold on to many nutrients, and they usually contain less clays. If your soil has a low CEC, it is important to apply fertilizer small doses so it does not infiltrate into the groundwater. A soil with a low CEC is less able to hold spilt chemicals.

Soil pH

The soil pH is a measure of soil acidity or alkalinity. pH can range from 1 to 14, with values 0-7 being acidic, and 7-14 being alkaline. Soils usually range from 4 to 10. The pH is one of the most important properties involved in plant growth, as well as understanding how rapidly reactions occur in the soil. .For example, the element iron becomes less available to plants a higher the pH is. This creates iron deficiency problems. Crops usually prefer values between 5.5-8, but the value depends on the crop. The pH of soil comes from the parent material during soil formation, but humans can add things to soils to change them to better suit plant growth. Soil pH also affects organisms.

Sorption and Precipitation

Soil particles have the ability to capture different nutrients and ions. Sorption is the process in which one substance takes up or holds another. In this case, soils that have high sorption can hold a lot of extra environmental contaminents, like phosporus, onto the particles. Soil precipitation occurs during chemical reactions when a nutrient or chemical in the soil solution (water around soil particles) transforms into a solid. This is really important if soils are really salty. Soil chemists study the speed of these reactions under many different conditions.

Soil Organic Matter Interactions

Soil chemists also study soil organic matter (OM), which are materials derived from the decay of plants and animals. They contain many hydrogen and carbon compounds. The arrangement and formation of these compounds influence a soils ability to handle spilt chemicals and other pollutants.

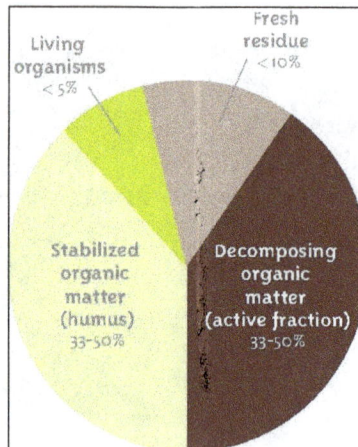

Soil has four major categories of organic matter inside of it, including active and long term types.

Oxidation and Reduction Reactions

Soils that alternate between wet and dry go from having a lot of oxygen to not a lot of oxygen. The presence or absence of oxygen determines how soils chemically react. Oxidation is the loss of electrons, and reduction is the gaining of electrons at the soil surface. These type of reactions occur every day, and are responsible for creating things like rust. Soils, because they contain a lot of iron, can also rust, or if they contain a lot of water, can turn a light gray color. This is partially responsible for all of the different colors that are found, and creates the speckles usually found deeper in the soil.

SOIL BIOLOGY

Soil biology is the study of microbial and faunal activity in the soil. This photo shows the activity of both.

Soil biology is the study of microbial and faunal activity and ecology in soil. Soil life, soil biota, soil fauna, or edaphon is a collective term that encompasses all organisms that spend a significant portion of their life cycle within a soil profile, or at the soil-litter interface. These organisms include earthworms, nematodes, protozoa, fungi, bacteria, different arthropods, as well as some reptiles (such as snakes), and species of burrowing mammals like gophers, moles and prairie dogs. Soil biology plays a vital role in determining many soil characteristics. The decomposition of organic matter by soil organisms has an immense influence on soil fertility, plant growth, soil structure, and carbon storage. As a relatively new science, much remains unknown about soil biology and its effect on soil ecosystems.

The soil is home to a large proportion of the world's biodiversity. The links between soil organisms and soil functions are observed to be incredibly complex. The interconnectedness and complexity of this soil 'food web' means any appraisal of soil function must necessarily take into account interactions with the living communities that exist within the soil. We know that soil organisms break down organic matter, making nutrients available for uptake by plants and other organisms. The nutrients stored in the bodies of soil organisms prevent nutrient loss by leaching. Microbial exudates act to maintain soil structure, and earthworms are important in bioturbation. However, we find that we don't understand critical aspects about how these populations function and interact. The discovery of glomalin in 1995 indicates that we lack the knowledge to correctly answer some of the most basic questions about the biogeochemical cycle in soils. There is much work ahead to gain a better understanding of the ecological role of soil biological components in the biosphere.

In balanced soil, plants grow in an active and steady environment. The mineral content of the soil and its profound structure are important for their well-being, but it is the life in the earth that powers its cycles and provides its fertility. Without the activities of soil organisms, organic materials would accumulate and litter the soil surface, and there would be no food for plants. The soil biota includes:

- Megafauna: Size range - 20 mm upward, e.g. moles, rabbits, and rodents.

- Macrofauna: Size range - 2 to 20 mm, e.g. woodlice, earthworms, beetles, centipedes, slugs, snails, ants, and harvestmen.

- Mesofauna: Size range - 100 micrometres to 2 mm, e.g. tardigrades, mites and springtails.

- Microfauna and Microflora: Size range - 1 to 100 micrometres, e.g. yeasts, bacteria (commonly actinobacteria), fungi, protozoa, roundworms, and rotifers.

Of these, bacteria and fungi play key roles in maintaining a healthy soil. They act as decomposers that break down organic materials to produce detritus and other breakdown products. Soil detritivores, like earthworms, ingest detritus and decompose it. Saprotrophs, well represented by fungi and bacteria, extract soluble nutrients from delitro. The ants (macrofaunas) help by breaking down in the same way but they also provide the motion part as they move in their armies. Also the rodents, wood-eaters help the soil to be more absorbent.

Scope

Soil biology involves work in the following areas:

- Modelling of biological processes and population dynamics.

- Soil biology, physics and chemistry: Occurrence of physicochemical parameters and surface properties on biological processes and population behavior.

- Population biology and molecular ecology: Methodological development and contribution to study microbial and faunal populations; diversity and population dynamics; genetic transfers, influence of environmental factors.

- Community ecology and functioning processes: Interactions between organisms and mineral or organic compounds; involvement of such interactions in soil pathogenicity; transformation of mineral and organic compounds, cycling of elements; soil structuration.

Complementary disciplinary approaches are necessarily utilized which involve molecular biology, genetics, ecophysiology, biogeography, ecology, soil processes, organic matter, nutrient dynamics and landscape ecology.

Bacteria

Bacteria are single-cell organisms and the most numerous denizens of agriculture, with populations ranging from 100 million to 3 billion in a gram. They are capable of very rapid reproduction by binary fission (dividing into two) in favourable conditions. One bacterium is capable of producing 16 million more in just 24 hours. Most soil bacteria live close to plant roots and are often referred to as rhizobacteria. Bacteria live in soil water, including the film of moisture surrounding soil particles, and some are able to swim by means of flagella. The majority of the beneficial soil-dwelling bacteria need oxygen (and are thus termed aerobic bacteria), whilst those that do not require air are referred to as anaerobic, and tend to cause putrefaction of dead organic matter. Aerobic bacteria are most active in a soil that is moist (but not saturated, as this will deprive aerobic bacteria of the air that they require), and neutral soil pH, and where there is plenty of food (carbohydrates and micronutrients from organic matter) available. Hostile conditions will not completely kill bacteria; rather, the bacteria will stop growing and get into a dormant stage, and those individuals with pro-adaptive mutations may compete better in the new conditions. Some gram-positive bacteria produce spores in order to wait for more favourable circumstances, and gram-negative bacteria get into a "nonculturable" stage. Bacteria are colonized by persistent viral agents (bacteriophages) that determine gene word order in bacterial host.

From the organic gardener's point of view, the important roles that bacteria play are:

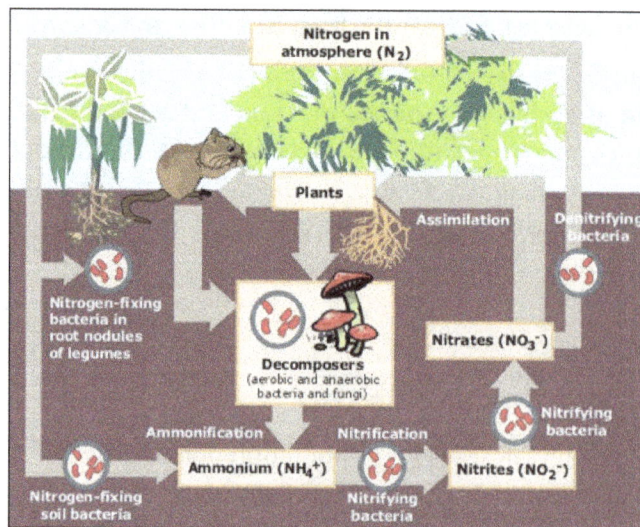

The nitrogen cycle.

Nitrification

Nitrification is a vital part of the nitrogen cycle, wherein certain bacteria (which manufacture their own carbohydrate supply without using the process of photosynthesis) are able to transform nitrogen in the form of ammonium, which is produced by the decomposition of proteins, into nitrates, which are available to growing plants, and once again converted to proteins.

Nitrogen Fixation

In another part of the cycle, the process of nitrogen fixation constantly puts additional nitrogen into biological circulation. This is carried out by free-living nitrogen-fixing bacteria in the soil or water such as *Azotobacter*, or by those that live in close symbiosis with leguminous plants, such as rhizobia. These bacteria form colonies in nodules they create on the roots of peas, beans, and related species. These are able to convert nitrogen from the atmosphere into nitrogen-containing organic substances.

Denitrification

While nitrogen fixation converts nitrogen from the atmosphere into organic compounds, a series of processes called denitrification returns an approximately equal amount of nitrogen to the atmosphere. Denitrifying bacteria tend to be anaerobes, or facultatively anaerobes (can alter between the oxygen dependent and oxygen independent types of metabolisms), including *Achromobacter* and *Pseudomonas*. The purification process caused by oxygen-free conditions converts nitrates and nitrites in soil into nitrogen gas or into gaseous compounds such as nitrous oxide or nitric oxide. In excess, denitrification can lead to overall losses of available soil nitrogen and subsequent loss of soil fertility. However, fixed nitrogen may circulate many times between organisms and the soil before denitrification returns it to the atmosphere.

Actinobacteria

Actinobacteria are critical in the decomposition of organic matter and in humus formation. They specialize in breaking down cellulose and lignin along with the tough chitin found on the exoskeletons of insects. Their presence is responsible for the sweet "earthy" aroma associated with a good healthy soil. They require plenty of air and a pH between 6.0 and 7.5, but are more tolerant of dry conditions than most other bacteria and fungi.

Fungi

A gram of garden soil can contain around one million fungi, such as yeasts and moulds. Fungi have no chlorophyll, and are not able to photosynthesise. They cannot use atmospheric carbon dioxide as a source of carbon, therefore they are chemo-heterotrophic, meaning that, like animals, they require a chemical source of energy rather than being able to use light as an energy source, as well as organic substrates to get carbon for growth and development.

Many fungi are parasitic, often causing disease to their living host plant, although some have beneficial relationships with living plants. In terms of soil and humus creation, the most important fungi tend to be saprotrophic; that is, they live on dead or decaying organic matter, thus breaking it down and converting it to forms that are available to the higher plants. A succession of fungi species will colonise the dead matter, beginning with those that use sugars and starches, which are succeeded by those that are able to break down cellulose and lignins.

Fungi spread underground by sending long thin threads known as mycelium throughout the soil; these threads can be observed throughout many soils and compost heaps. From the mycelia the fungi is able to throw up its fruiting bodies, the visible part above the soil (e.g., mushrooms, toadstools, and puffballs), which may contain millions of spores. When the fruiting body bursts, these spores are dispersed through the air to settle in fresh environments, and are able to lie dormant for up to years until the right conditions for their activation arise or the right food is made available.

Mycorrhizae

Those fungi that are able to live symbiotically with living plants, creating a relationship that is beneficial to both, are known as Mycorrhizae (from *myco* meaning fungal and *rhiza* meaning root). Plant root hairs are invaded by the mycelia of the mycorrhiza, which lives partly in the soil and partly in the root, and may either cover the length of the root hair as a sheath or be concentrated around its tip. The mycorrhiza obtains the carbohydrates that it requires from the root, in return providing the plant with nutrients including nitrogen and moisture. Later the plant roots will also absorb the mycelium into its own tissues.

Beneficial mycorrhizal associations are to be found in many of our edible and flowering crops. Shewell Cooper suggests that these include at least 80% of the *brassica* and *solanum* families (including tomatoes and potatoes), as well as the majority of tree species, especially in forest and woodlands. Here the mycorrhizae create a fine underground mesh that extends greatly beyond the limits of the tree's roots, greatly increasing their feeding range and actually causing neighbouring trees to become physically interconnected. The benefits of mycorrhizal relations to their plant partners are not limited to nutrients, but can be essential for plant reproduction. In situations where little light is able to reach the forest floor, such as the North American pine forests, a young seedling cannot obtain sufficient light to photosynthesise for itself and will not grow properly in a sterile soil. But, if the ground is underlain by a mycorrhizal mat, then the developing seedling will throw down roots that can link with the fungal threads and through them obtain the nutrients it needs, often indirectly obtained from its parents or neighbouring trees.

David Attenborough points out the plant, fungi, animal relationship that creates a "Three way harmonious trio" to be found in forest ecosystems, wherein the plant/fungi symbiosis is enhanced by animals such as the wild boar, deer, mice, or flying squirrel, which feed upon the fungi's fruiting bodies, including truffles, and cause their further spread. A greater understanding of the complex relationships that pervade natural systems is one of the major justifications of the organic gardener, in refraining from the use of artificial chemicals and the damage these might cause.

Recent research has shown that arbuscular mycorrhizal fungi produce glomalin, a protein that binds soil particles and stores both carbon and nitrogen. These glomalin-related soil proteins are an important part of soil organic matter.

Insects and Mammals in Soil

Earthworms, ants and termites mix the soil as they burrow, significantly affecting soil formation. Earthworms ingest soil particles and organic residues, enhancing the availability of plant nutrients in the material that passes through and out of their bodies. By aerating and stirring the soil, and by increasing the stability of soil aggregates, these organisms help to assure the ready infiltration of water. These organisms in the soil also help improve your pH levels.

Ants and termites are often referred to as "Soil engineers" as when they create their nests there are several chemical and physical changes that become present. Such as increasing the presence of the most essential elements like Carbon , Nitrogen and phosphorus. Which are all essential for plant and other organisms growth. They also can gather soil particles from differing depth levels of soil and deposit them in other places. Which then leads to the mixing of soil so it is richer with nutrients and other elements.

Gopher sticking out of burrow.

Many mammals also find the soil to be very important to them. As many mammals such as gophers, moles , prairie dogs and other burrowing animals rely on this soil for protection and food. The animals even give back to the soil as them burrowing allows more rain , snow and water from ice to enter the soil instead of creating erosion.

SOIL MICROBIOLOGY

Soil microbiology is the study of microorganisms in soil, their functions, and how they affect soil properties. It is believed that between two and four billion years ago, the first ancient bacteria and microorganisms came about on Earth's oceans. These bacteria could fix nitrogen, in time multiplied, and as a result released oxygen into the atmosphere. This led to more advanced microorganisms, which are important because they affect soil structure and fertility. Soil microorganisms can be classified as bacteria, actinomycetes, fungi, algae and protozoa. Each of these groups has characteristics that define them and their functions in soil.

Up to 10 billion bacterial cells inhabit each gram of soil in and around plant roots, a region known as the rhizosphere. In 2011, a team detected more than 33,000 bacterial and archaeal species on sugar beet roots.

The composition of the rhizobiome can change rapidly in response to changes in the surrounding environment.

Bacteria

Bacteria and Archaea are the smallest organisms in soil apart from viruses. Bacteria and Archaea are prokaryotic. All of the other microorganisms are eukaryotic, which means they have a more advanced cell structure with internal organelles and the ability to reproduce sexually. A prokaryote has a very simple cell structure with no internal organelles. Bacteria and archaea are the most abundant microorganisms in the soil, and serve many important purposes, including nitrogen fixation.

Biochemical Processes

One of the most distinguished features of bacteria is their biochemical versatility . A bacterial genus called *Pseudomonas* can metabolize a wide range of chemicals and fertilizers. In contrast, another genus known as *Nitrobacter* can only derive its energy by turning nitrite into nitrate, which is also known as oxidation. The genus *Clostridium* is an example of bacterial versatility because it, unlike most species, can grow in the absence of oxygen, respiring anaerobically. Several species of *Pseudomonas*, such as *Pseudomonas aeruginosa* are able to respire both aerobically and anaerobically, using nitrate as the terminal electron acceptor.

Nitrogen Fixation

Bacteria are responsible for the process of nitrogen fixation, which is the conversion of atmospheric nitrogen into nitrogen-containing compounds (such as ammonia) that can be used by plants. Autotrophic bacteria derive their energy by making their own food through oxidation, like the *Nitrobacters* species, rather than feeding on plants or other organisms. These bacteria are responsible for nitrogen fixation. The amount of autotrophic bacteria is small compared to heterotrophic bacteria (the opposite of autotrophic bacteria, heterotrophic bacteria acquire energy by consuming plants or other microorganisms), but are very important because almost every plant and organism requires nitrogen in some way.

Actinomycetes

Actinomycetes are soil microorganisms. They are a type of bacteria, but they share some characteristics with fungi that are most likely a result of convergent evolution due to a common habitat and lifestyle.

Similarities to Fungi

Although they are members of the Bacteria kingdom, many actinomycetes share characteristics with fungi, including shape and branching properties, spore formation and secondary metabolite production.

- The mycelium branches in a manner similar to that of fungi.
- They form aerial mycelium as well as conidia.
- Their growth in liquid culture occurs as distinct clumps or pellets, rather than as a uniform turbid suspension as in bacteria.

Antibiotics

One of the most notable characteristics of the actinomycetes is their ability to produce antibiotics. Streptomycin, neomycin, erythromycinand tetracycline are only a few examples of these antibiotics. Streptomycin is used to treat tuberculosis and infections caused by certain bacteria and neomycin is used to reduce the risk of bacterial infection during surgery. Erythromycin is used to treat certain infections caused by bacteria, such as bronchitis, pertussis (whooping cough), pneumonia and ear, intestine, lung, urinary tract and skin infections.

Fungi

Fungi are abundant in soil, but bacteria are more abundant. Fungi are important in the soil as food sources for other, larger organisms, pathogens, beneficial symbiotic relationships with plants or other organisms and soil health. Fungi can be split into species based primarily on the size, shape and color of their reproductive spores, which are used to reproduce. Most of the environmental factors that influence the growth and distribution of bacteria and actinomycetes also influence fungi. The quality as well as quantity of organic matter in the soil has a direct correlation to the growth of fungi, because most fungi consume organic matter for nutrition. Fungi thrive in acidic environments, while bacteria and actinomycetes cannot survive in acid, which results in an abundance of fungi in acidic areas. Fungi also grow well in dry, arid soils because fungi are aerobic, or dependent on oxygen, and the higher the moisture content in the soil, the less oxygen is present for them.

Algae

Algae can make their own nutrients through photosynthesis. Photosynthesis converts light energy to chemical energy that can be stored as nutrients. For algae to grow, they must be exposed to light because photosynthesis requires light, so algae are typically distributed evenly wherever sunlight and moderate moisture is available. Algae do not have to be directly exposed to the Sun, but can live below the soil surface given uniform temperature and moisture conditions. Algae are also capable of performing nitrogen fixation.

Types

Algae can be split up into three main groups: the Cyanophyceae, the Chlorophyceae and the Bacillariaceae. The Cyanophyceae contain chlorophyll, which is the molecule that absorbs sunlight and uses that energy to make carbohydrates from carbon dioxide and water and also pigments that make it blue-green to violet in color. The Chlorophyceae usually only have chlorophyll in them which makes them green, and the Bacillariaceae contain chlorophyll as well as pigments that make the algae brown in color.

Blue-green Algae and Nitrogen Fixation

Blue-green algae, or Cyanophyceae, are responsible for nitrogen fixation. The amount of nitrogen they fix depends more on physiological and environmental factors rather than the organism's abilities. These factors include intensity of sunlight, concentration of inorganic and organic nitrogen sources and ambient temperature and stability.

Protozoa

Protozoa are eukaryotic organisms that were some of the first microorganisms to reproduce sexually, a significant evolutionary step from duplication of spores, like those that many other soil microorganisms depend on. Protozoa can be split up into three categories: flagellates, amoebae and ciliates.

Flagellates

Flagellates are the smallest members of the protozoa group, and can be divided further based on whether they can participate in photosynthesis. Nonchlorophyll-containing flagellates are not capable of photosynthesis because chlorophyll is the green pigment that absorbs sunlight. These flagellates are found mostly in soil. Flagellates that contain chlorophyll typically occur in aquatic conditions. Flagellates can be distinguished by their flagella, which is their means of movement. Some have several flagella, while other species only have one that resembles a long branch or appendage.

Amoebae

Amoebae are larger than flagellates and move in a different way. Amoebae can be distinguished from other protozoa by their slug-like properties and pseudopodia. A pseudopodium or "false foot" is a temporary obtrusion from the body of the amoeba that helps pull it along surfaces for movement or helps to pull in food. The amoeba does not have permanent appendages and the pseudopodium is more of a slime-like consistency than a flagellum.

Ciliates

Ciliates are the largest of the protozoa group, and move by means of short, numerous cilia that produce beating movements. Cilia resemble small, short hairs. They can move in different directions to move the organism, giving it more mobility than flagellates or amoebae.

Composition Regulation

Plant hormones salicylic acid, jasmonic acid and ethylene are key regulators of innate immunity in plant leaves. Mutants impaired in salicylic acid synthesis and signaling are hypersusceptible to microbes that colonize the host plant to obtain nutrients, whereas mutants impaired in jasmonic acid and ethylene synthesis and signaling are hypersusceptible to herbivorous insects and microbes that kill host cells to extract nutrients. The challenge of modulating a community of diverse microbes in plant roots is more involved than that of clearing a few pathogens from inside a plant leaf. Consequently, regulating root microbiome composition may require immune mechanisms other than those that control foliar microbes.

A 2015 study analyzed a panel of *Arabidopsis* hormone mutants impaired in synthesis or signaling of individual or combinations of plant hormones, the microbial community in the soil adjacent to the root and in bacteria living within root tissue. Changes in salicylic acid signaling stimulated a reproducible shift in the relative abundance of bacterial phyla in the endophytic compartment. These changes were consistent across many families within the affected phyla, indicating that salicylic acid may be a key regulator of microbiome community structure.

Classical plant defense hormones also function in plant growth, metabolism and abiotic stress responses, obscuring the precise mechanism by which salicylic acid regulates this microbiome.

During plant domestication, humans selected for traits related to plant improvement, but not for plant associations with a beneficial microbiome. Even minor changes in abundance of certain bacteria can have a major effect on plant defenses and physiology, with only minimal effects on overall microbiome structure.

Applications

Agriculture

Microbes can make nutrients and minerals in the soil available to plants, produce hormones that spur growth, stimulate the plant immune system and trigger or dampen stress responses. In general a more diverse soil microbiome results in fewer plant diseases and higher yield.

Farming can destroy soil's rhiziobiome (microbial ecosystem) by using soil amendments such as fertilizer and pesticide without compensating for their effects. By contrast, healthy soil can increase fertility in multiple ways, including supplying nutrients such as nitrogen and protecting against pests and disease, while reducing the need for water and other inputs. Some approaches may even allow agriculture in soils that were never considered viable.

The group of bacteria called rhizobia live inside the roots of legumes and fix nitrogen from the air into a biologically useful form.

Mycorrhizae or root fungi form a dense network of thin filaments that reach far into the soil, acting as extensions of the plant roots they live on or in. These fungi facilitate the uptake of water and a wide range of nutrients.

Up to 30% of the carbon fixed by plants is excreted from the roots as so-called exudates—including sugars, amino acids, flavonoids, aliphatic acids, and fatty acids—that attract and feed beneficial microbial species while repelling and killing harmful ones.

Commercial Activity

Almost all registered microbes are biopesticides, producing some $1 billion annually, less than 1% of the chemical amendment market, estimated at $110 billion. Some microbes have been marketed for decades, such as *Trichoderma* fungi that suppress other, pathogenic fungi, and the caterpillar killer *Bacillus thuringiensis*. Serenade is a biopesticide containing a *Bacillus subtilis* strain that has antifungal and antibacterial properties and promotes plant growth. It can be applied in a liquid form on plants and to soil to fight a range of pathogens. It has found acceptance in both conventional and organic agriculture.

Agrochemical companies such as Bayer have begun investing in the technology. In 2012, Bayer bought AgraQuest for $425 million. Its €10 million annual research budget funds field-tests of dozens of new fungi and bacteria to replace chemical pesticides or to serve as biostimulants to promote crop health and growth. Novozymes, a company developing microbial fertilizers and pesticides, forged an alliance with Monsanto. Novozymes invested in a biofertilizer containing the soil fungus *Penicillium bilaiae* and a bioinsecticide that contains the fungus *Metarhizium anisopliae*. In 2014 Syngenta and BASF acquired companies developing microbial products, as did Dupont in 2015.

A 2007 study showed that a complex symbiosis with fungi and viruses makes it possible for a grass called *Dichanthelium lanuginosum* to thrive in geothermal soils in Yellowstone National Park, where temperatures reach 60 °C (140 °F). Introduced in the US market in 2014 for corn and rice, they trigger an adaptive stress response.

In both the US and Europe, companies have to provide regulatory authorities with evidence that both the individual strains and the product as a whole are safe, leading many existing products to label themselves "biostimulants" instead of "biopesticides".

Unhelpful Microbes

A funguslike unicellular organism named *Phytophthora infestans*, responsible for potato blight and other crop diseases, has caused famines throughout history. Other fungi and bacteria cause the decay of roots and leaves.

Many strains that seemed promising in the lab often failed to prove effective in the field, because of soil, climate and ecosystem effects, leading companies to skip the lab phase and emphasize field tests.

Fade

Populations of beneficial microbes can diminish over time. Serenade stimulates a high initial *B. subtilis* density, but levels decrease because the bacteria lacks a defensible niche. One way to compensate is to use multiple collaborating strains.

Fertilizers deplete soil of organic matter and trace elements, cause salination and suppress mycorrhizae; they can also turn symbiotic bacteria into competitors.

Pilot Project

A pilot project in Europe used a plow to slightly loosen and ridge the soil. They planted oats and vetch, which attracts nitrogen-fixing bacteria. They planted small olive trees to boost microbial diversity. They split an unirrigated 100-hectare field into three zones, one treated with chemical fertilizer and pesticides; and the other two with different amounts of an organic biofertilizer, consisting of fermented grape leftovers and a variety of bacteria and fungi, along with four types of mycorrhiza spores.

The crops that had received the most organic fertilizer had reached nearly twice the height of those in zone A and were inches taller than zone C. The yield of that section equaled that of irrigated crops, whereas the yield of the conventional technique was negligible. The mycorrhiza had penetrated the rock by excreting acids, allowing plant roots to reach almost 2 meters into the rocky soil and reach groundwater.

SOIL ECOLOGY

Soil ecology is the study of the interactions among soil biology, and between biotic and abiotic aspects of the soil environment. It is particularly concerned with the cycling of nutrients, formation and stabilization of the pore structure, the spread and vitality of pathogens, and the biodiversity of this rich biological community.

Soil is made up of a multitude of physical, chemical, and biological entities, with many interactions occurring among them. Soil is a variable mixture of broken and weathered minerals and decaying organic matter. Together with the proper amounts of air and water, it supplies, in part, sustenance for plants as well as mechanical support.

The diversity and abundance of soil life exceeds that of any other ecosystem. Plant establishment, competitiveness, and growth is governed largely by the ecology below-ground, so understanding this system is an essential component of plant sciences and terrestrial ecology.

Features of the Ecosystem

- Moisture is a major limiting factor on land. Terrestrial organisms are constantly confronted with the problem of dehydration. Transpiration or evaporation of water from plant surfaces is an energy dissipating process unique to the terrestrial environment.

- Temperature variations and extremes are more pronounced in the air than in the water medium.

- On the other hand, the rapid circulation of air throughout the globe results in a ready mixing and remarkably constant content of oxygen and carbon dioxide.

- Although soil offers solid support, air does not. Storing skeletons have been evolved in both land plants and animals and also special means of locomotion have been evolved in the latter.

- Land, unlike the ocean, is not continuous; there are important geographical barriers to free movement.

- The nature of the substrate, although important in water is especially vital in terrestrial environment. Soil, not air, is the source of highly variable nutrients; it is a highly developed ecological subsystem.

Soil Food Web

An incredible diversity of organisms make up the soil food web. They range in size from the tiniest one-celled bacteria, algae, fungi, and protozoa, to the more complex nematodes and micro-arthropods, to the visible earthworms, insects, small vertebrates, and plants. As these organisms eat, grow, and move through the soil, they make it possible to have clean water, clean air, healthy plants, and moderated water flow.

There are many ways that the soil food web is an integral part of landscape processes. Soil organisms decompose organic compounds, including manure, plant residues, and pesticides, preventing them from entering water and becoming pollutants. They sequester nitrogen and other nutrients that might otherwise enter groundwater, and they fix nitrogen from the atmosphere, making it available to plants. Many organisms enhance soil aggregation and porosity, thus increasing infiltration and reducing surface runoff. Soil organisms prey on crop pests and are food for above-ground animals.

PEDOLOGY

Soil Profile on Chalk at Seven Sisters Country Park, England.

Pedology is the study of soils in their natural environment. It is one of two main branches of soil science, the other being edaphology. Pedology deals with pedogenesis, soil morphology, and soil classification, while edaphology studies the way soils influence plants, fungi, and other living things. The quantitative branch of pedology is called pedometrics.

Soil is not only a support for vegetation, but it is also the pedosphere, the locus of numerous interactions between climate (water, air, temperature), soil life (micro-organisms, plants, animals) and its residues, the mineral material of the original and added rock, and its position in the landscape. During its formation and genesis, the soil profile slowly deepens and develops characteristic layers, called 'horizons', while a steady state balance is approached.

Soil users (such as agronomists) showed initially little concern in the dynamics of soil. They saw it as medium whose chemical, physical and biological properties were useful for the services of agronomic productivity. On the other hand, pedologists and geologists did not initially focus on the agronomic applications of the soil characteristics (edaphic properties) but upon its relation to the nature and history of landscapes. Today, there is an integration of the two disciplinary approaches as part of landscape and environmental sciences.

Pedologists are now also interested in the practical applications of a good understanding of pedogenesis processes (the evolution and functioning of soils), like interpreting its environmental history and predicting consequences of changes in land use, while agronomists understand that the cultivated soil is a complex medium, often resulting from several thousands of years of evolution. They understand that the current balance is fragile and that only a thorough knowledge of its history makes it possible to ensure its sustainable use.

Concepts

Important pedological concepts include:

- Complexity in soil genesis is more common than simplicity.

- Soils lie at the interface of Earth's atmosphere, biosphere, hydrosphere and lithosphere. Therefore, a thorough understanding of soils requires some knowledge of meteorology, climatology, ecology, biology, hydrology, geomorphology, geology and many other earth sciences and natural sciences.

- Contemporary soils carry imprints of pedogenic processes that were active in the past, although in many cases these imprints are difficult to observe or quantify. Thus, knowledge of paleoecology, palaeogeography, glacial geology and paleoclimatology is important for the recognition and understanding of soil genesis and constitute a basis for predicting future soil changes.

- Five major, external factors of formation (climate, organisms, relief, parent material and time), and several smaller, less identifiable ones, drive pedogenic processes and create soil patterns.

- Characteristics of soils and soil landscapes, e.g., the number, sizes, shapes and arrangements of soil bodies, each of which is characterized on the basis of soil horizons, degree of internal homogeneity, slope, aspect, landscape position, age and other properties and relationships, can be observed and measured.

- Distinctive bioclimatic regimes or combinations of pedogenic processes produce distinctive soils. Thus, distinctive, observable morphological features, e.g., illuvial clay accumulation in B horizons, are produced by certain combinations of pedogenic processes operative over varying periods of time.

- Pedogenic (soil-forming) processes act to both create and destroy order (anisotropy) within soils; these processes can proceed simultaneously. The resulting soil profile reflects the balance of these processes, present and past.

- The geological Principle of Uniformitarianism applies to soils, i.e., pedogenic processes active in soils today have been operating for long periods of time, back to the time of appearance of organisms on the land surface. These processes do, however, have varying degrees of expression and intensity over space and time.

- A succession of different soils may have developed, eroded and/or regressed at any particular site, as soil genetic factors and site factors, e.g., vegetation, sedimentation, geomorphology, change.

- There are very few old soils (in a geological sense) because they can be destroyed or buried by geological events, or modified by shifts in climate by virtue of their vulnerable position at the surface of the earth. Little of the soil continuum dates back beyond the Tertiaryperiod and most soils and land surfaces are no older than the Pleistocene Epoch. However, preserved/lithified soils (paleosols) are an almost ubiquitous feature in terrestrial (land-based) environments throughout most of geologic time. Since they record evidence of ancient climate change, they present immense utility in understanding climate evolution throughout geologic history.

- Knowledge and understanding of the genesis of a soil is important in its classification and mapping.

- Soil classification systems cannot be based entirely on perceptions of genesis, however, because genetic processes are seldom observed and because pedogenic processes change over time.

- Knowledge of soil genesis is imperative and basic to soil use and management. Human influence on, or adjustment to, the factors and processes of soil formation can be best controlled and planned using knowledge about soil genesis.

- Soils are natural clay factories (clay includes both clay mineral structures and particles less than 2 µm in diameter). Shales worldwide are, to a considerable extent, simply soil clays that have been formed in the pedosphere and eroded and deposited in the ocean basins, to become lithified at a later date.

EDAPHOLOGY

Edaphology is one of two main divisions of soil science, the other being pedology. Edaphology is concerned with the influence of soils on living things, particularly plants. Edaphology includes the study of how soil influences humankind's use of land for plant growth as well as man's overall use of the land. General subfields within edaphology are agricultural soil science (known by the term agrology in some regions) and environmental soil science. (Pedology deals with pedogenesis, soil morphology, and soil classification.).

In Russia, edaphology is considered equivalent to pedology, but is recognized to have an applied sense consistent with agrophysics and agrochemistry outside Russia.

Areas of Study

Agricultural soil science is the application of soil chemistry, physics, and biology dealing with the production of crops. In terms of soil chemistry, it places particular emphasis on plant nutrients of importance to farming and horticulture, especially with regard to soil fertilityand fertilizer components.

Physical edaphology is strongly associated with crop irrigation and drainage.

Soil husbandry is a strong tradition within agricultural soil science. Beyond preventing soil erosion and degradation in cropland, soil husbandry seeks to sustain the agricultural soil resource though the use of soil conditioners and cover crops.

Environmental Soil Science

Environmental soil science studies our interaction with the pedosphere on beyond crop production. Fundamental and applied aspects of the field address vadose zone functions, septic drain field site assessment and function, land treatment of wastewater, stormwater, erosioncontrol, soil contamination with metals and pesticides, remediation of contaminated soils, restoration of wetlands, soil degradation, and environmental nutrient management. It also studies soil in the context of land-use planning, global warming and acid rain.

Agricultural Soil Science

Agricultural soil science is a branch of soil science that deals with the study of edaphicconditions as they relate to the production of food and fiber. In this context, it is also a constituent of the field of agronomy.

Agricultural soil science follows the holistic method. Soil is investigated in relation to and as integral part of terrestrial ecosystems but is also recognized as a manageable natural resource.

Agricultural soil science studies the chemical, physical, biological, and mineralogical composition of soils as they relate to agriculture. Agricultural soil scientists develop methods that will improve the use of soil and increase the production of food and fiber crops. Emphasis continues to grow on the importance of soil sustainability. Soil degradation such as erosion, compaction, lowered fertility, and contamination continue to be serious concerns. They conduct research in irrigation and drainage, tillage, soil classification, plant nutrition, soil fertility and other areas.

Although maximizing plant (and thus animal) production is a valid goal, sometimes it may come at high cost which can be readily evident (e.g. massive crop disease stemming from monoculture) or long-term (e.g. impact of chemical fertilizers and pesticides on human health). An agricultural soil scientist may come up with a plan that can maximize production using sustainable methods and solutions, and in order to do that he must look into a number of science fields including agricultural science, physics, chemistry, biology, meteorology and geology.

Kinds of Soil and their Variables

Some soil variables of special interest to agricultural soil science are:

- Soil texture or soil composition: Soils are composed of solid particles of various sizes. In decreasing order, these particles are sand, siltand clay. Every soil can be classified according to the relative percentage of sand, silt and clay it contains.

- Aeration and porosity: Atmospheric air contains elements such as oxygen, nitrogen, carbon and others. These elements are prerequisites for life on Earth. Particularly, all cells (including root cells) require oxygen to function and if conditions become anaerobicthey fail to respire and metabolize. Aeration in this context refers to the mechanisms by which air is delivered to the soil. In natural ecosystems soil aeration is chiefly accomplished through the vibrant activity of the biota. Humans commonly aerate the soil by tilling and plowing, yet such practice may cause degradation. Porosity refers to the air-holding capacity of the soil.

- Drainage: In soils of bad drainage the water delivered through rain or irrigation may pool and stagnate. As a result, prevail anaerobicconditions and plant roots suffocate. Stagnant water also favors plant-attacking water molds. In soils of excess drainage, on the other hand, plants don't get to absorb adequate water and nutrients are washed from the porous medium to end up in groundwater reserves.

- Water content: Without soil moisture there is no transpiration, no growth and plants wilt. Technically, plant cells lose their pressure. Plants contribute directly to soil moisture. For instance, they create a leafy cover that minimizes the evaporative effects of solar radiation. But even when plants or parts of plants die, the decaying plant matter produces a thick organic cover that protects the soil from evaporation, erosion and compaction.

- Water potential: Water potential describes the tendency of the water to flow from one area of the soil to another. While water delivered to the soil surface normally flows downward due to gravity, at some point it meets increased pressure which causes a reverse upward flow. This effect is known as water suction.

- Horizonation: Typically found in advanced and mature soils, horizonation refers to the creation of soil layers with differing characteristics. It affects almost all soil variables.

- Fertility: A fertile soil is one rich in nutrients and organic matter. Modern agricultural methods have rendered much of the arable land infertile. In such cases, soil can no longer support on its own plants with high nutritional demand and thus needs an external source of nutrients. However, there are cases where human activity is thought to be responsible for transforming rather normal soils into super-fertile ones.

- Biota and soil biota: Organisms interact with the soil and contribute to its quality in innumerable ways. Sometimes the nature of interaction may be unclear, yet a rule is becoming evident: The amount and diversity of the biota is "proportional" to the quality of the soil. Clades of interest include bacteria, fungi, nematodes, annelids and arthropods.

- Soil acidity or soil pH and cation-exchange capacity: Root cells act as hydrogen pumps and the surrounding concentration of hydrogen ions affects their ability to absorb nutrients. pH is a measure of this concentration. Each plant species achieves maximum growth in a particular pH range, yet the vast majority of edible plants can grow in soil pH between 5.0 and 7.5.

Soil scientists use a soil classification system to describe soil qualities. The International Union of Soil Sciences endorses the World Reference Base as the international standard.

Soil Fertility

Agricultural soil scientists study ways to make soils more productive. They classify soils and test them to determine whether they contain nutrients vital to plant growth. Such nutritional substances include compounds of nitrogen, phosphorus, and potassium. If a certain soil is deficient in these substances, fertilizers may provide them. Agricultural soil scientists investigate the movement of nutrients through the soil, and the amount of nutrients absorbed by a plant's roots. Agricultural soil scientists also examine the development of roots and their relation to the soil. Some agricultural soil scientists try to understand the structure and function of soils in relation to soil fertility. They grasp the structure of soil as porous solid. The solid frames of soil consist of mineral derived from the rocks and organic matter originated from the dead bodies of various organisms. The pore space of the soil is essential for the soil to become productive. Small pores serve as water reservoir supplying water to plants and other organisms in the soil during the rain-less period. The water in the small pores of soils is not pure water; they call it soil solution. In soil solution, various plant nutrients derived from minerals and organic matters in the soil are there. This is measured through the cation exchange capacity. Large pores serve as water drainage pipe to allow the excessive water pass through the soil, during the heavy rains. They also serve as air tank to supply oxygen to plant roots and other living beings in the soil. In short, agricultural soil scientists see the soil as a vessel, the most precious one for us, containing all of the substances needed by the plants and other living beings on earth.

Soil Preservation

In addition, agricultural soil scientists develop methods to preserve the agricultural productivity of soil and to decrease the effects on productivity of erosion by wind and water. For example, a technique called contour plowing may be used to prevent soil erosion and conserve rainfall. Researchers in agricultural soil science also seek ways to use the soil more effectively in addressing associated challenges. Such challenges include the beneficial reuse of human and animal wastes using agricultural crops; agricultural soil management aspects of preventing water pollution and the build-up in agricultural soil of chemical pesticides. Regenerative agriculturepractices can be used to address these challenges and rebuild soil health.

Environmental Soil Science

Environmental soil science is the study of the interaction of humans with the pedosphere as well as critical aspects of the biosphere, the lithosphere, the hydrosphere, and the atmosphere. Environmental soil science addresses both the fundamental and applied aspects of the field including: buffers and surface water quality, vadose zone functions, septic drain field site assessment and function, land treatment of wastewater, stormwater, erosion control, soil contamination with metals and pesticides, remediation of contaminated soils, restoration of wetlands, soil degradation, nutrient management, movement of viruses and bacteria in soils and waters, bioremediation, application of molecular biology and genetic engineering to development of soil microbes that can degrade hazardous pollutants, land use, global warming, acid rain, and the study of anthropogenic soils, such as terra preta. Much of the research done in environmental soil science is produced through the use of models.

References

- Rca, nra, technical, detail, nrcs portal, wps: nrcs.usda.govm, Retrieved 9 April, 2019

- Sumner, Malcolm E.; Yamada, Tsuioshi (November 2002). "Farming with acidity". Communications in Soil Science and Plant Analysis. 33 (15–18): 2467–2496. Doi:10.1081/CSS-120014461

- Crop Tolerance to Soil Salinity, Statistical Analysis of Data Measured in Farm Lands. In: International Journal of Agricultural Science, October 2018. On line

- Henning Steinfeld; Pierre Gerber; Tom Wassenaar; Vincent Castel; Mauricio Rosales; Cees de Haan (2006). "Livestock's Long Shadow: Environmental issues and options". Food and Agriculture Organization of the United Nations. Retrieved 25 October 2012

- Soil-organic-matter, soil-management, land-management-and-soils, agriculture: dpipwe.tas.gov.au, Retrieved 10 May, 2019

- Drainage Manual: A Guide to Integrating Plant, Soil, and Water Relationships for Drainage of Irrigated Lands, Interior Dept., Bureau of Reclamation, 1993, ISBN 978-0-16-061623-5

Physical Properties of Soil

Some of the aspects that come under the physical properties of soil are soil texture, soil color, soil structure, soil porosity, soil density, soil texture and soil resistivity. This chapter has been carefully written to provide an easy understanding of these various physical properties of soil.

SOIL TEXTURE

Soil texture refers to the relative percentage of sand, silt and clay in a soil. Natural soils are comprised of soil particles of varying sizes. Texture is an important soil characteristic because it will partly determine water intake rates (absorption), water storage in the soil, and the ease of tillage operation, aeration status etc. and combinedly influence soil fertility.

As for an example, a coarse sandy soil is easy to cultivate or till, has sufficient aeration for good root growth and is easily wetted, but it also dries rapidly and easily loses plant nutrients through leaching. Whereas in case of high-clay soils (> 35% clay) have very small particles that fit tightly together, leaving very little pore spaces which permits very little room for water to flow into the soil. This condition makes soils difficult to wet, drain and till.

As the soil is a mixture of various sizes of soil separates, it is therefore, necessary to establish limits of variation for the soil separates with a view to group them into different textural classes. Texture is a basic property of a soil and it cannot be altered or changed.

Classes of Soil Texture

Texture names are given to soils based upon the relative proportion of each of the three soil categories such as sand, silt and clay. Soil that are preponderantly clay, are called clay (textural class), those with high silt content are silt (textural class) those with high sand percentage are sand (textural class). Three broad and fundamental groups of soil texture classes are recognised: sands, loams and clays.

- Sands: The sand group includes all soils of which the sand separates make up 70 per cent or more of the material by weight. Two specific classes are recognised sand and loamy sand.

- Loams: A loamy soil containing many sub-divisions does not exhibit the dominant physical properties of any of these three soils separates sand, silt and clay. An ideal loam soil may be defined as a mixture of sand, silt and clay particles which exhibits light and heavy properties in about equal proportions. Loam does not contain equal percentages of sand, silt and clay. It does, however, exhibit approximately equal properties of sand, silt and clay.

- Clay: A clay soil must carry at least 35 per cent of the clay separate and in most cases not less than 40 per cent. For an example, sandy clay soils contain more sand than clay. Similarly silty clay soils contain more silt than that of the clay. Based on these three broad and fundamental groups, the different textural class names developed by U.S. Department of Agriculture and U.S. Bureau of soils are presented in table.

Common name	Texture	Basic soil textural class name
Sandy soils	Coarse ⟶	Sandy Loamy sands Sandy loam
Loamy soils	Moderately coarse ⟶	Fine sandy loam Very fine sandy loam
	Medium ⟶	Loam Silt loam Silt
	Moderately fine ⟶	Clay loam Sandy clay loam Silty clay loam
Clayey soils	Fine ⟶	Sandy clay Silty clay Clay

Textural Class Names Developed by U.S Department of Agriculture.

Table: Textural Groups on the Basis of Percentages of Sand, Silt and Clay Separates.

	Textural group	Sand	Silt	Clay
1	Sand	80-100	0-20	0-20
2	Loam	50-80	0-50	0-20
3	Sandy loam	30-50	30-50	0-20
4	Silt loam	0-50	50-100	0-20
5	Sandy clay loam	50-80	0-30	20-30
6	Silty clay loam	0-30	50-80	20-30
7	Clay loam	20-50	20-50	20-30
8	Sandy clay	50-70	0-20	30-50
9	Silty clay	0-20	50-70	30-50
10	Clay	0-50	0-50	30-100

Determination of Soil Texture

- Feel Method: In the field, texture is commonly determined by the sense of feel. The soil is rubbed between thumb and fingers under wet conditions. Sands feel gritty and its particles can be easily seen. The silt when dry feels like flour and talcum powder and is slightly plastic when wet. Clayey particles feel very plastic and exhibit stickiness when wet and are hard under dry conditions.

- Laboratory Method: A more accurate and fundamental method has been devised by the U.S.

Department of Agriculture for the naming of soils based on a mechanical analysis. From the figure, mentioned earlier the determination of textural class names can be easily made.

SOIL COLOR

Soil color does not affect the behavior and use of soil; however, it can indicate the composition of the soil and give clues to the conditions that the soil is subjected to. Soil can exhibit a wide range of color; grey, black, white, reds, browns, yellows and under the right conditions green. Varying horizontal bands of color in the soil often identify a specific soil horizon. The development and distribution of color in soil results from chemical and biological weathering, especially redox reactions. As the primary minerals in soil parent material weather, the elements combine into new and colorful compounds. Soil conditions produce uniform or gradual color changes, while reducing environments result in disrupted color flow with complex, mottled patterns and points of color concentration.

Causes

Soil color is produced by the minerals present and by the organic matter content. Yellow or red soil indicates the presence of oxidized ferriciron oxides. Dark brown or black color in soil indicates that the soil has a high organic matter content. Wet soil will appear darker than dry soil. However, the presence of water also affects soil color by affecting the oxidation rate. Soil that has a high water content will have less air in the soil, specifically less oxygen. In well drained (and therefore oxygen rich soils) red and brown colors caused by oxidation are more common, as opposed to in wet (low oxygen) soils where the soil usually appears grey or greenish due to the presence of reduced (ferrous) iron oxide. The presence of other minerals can also affect soil color. Manganese oxide causes a black color, glauconite makes the soil green and calcite can make soil in arid regions appear white.

Organic matter tends to make the soil color darker. Humus, the final stage of organic matter breakdown is black. Throughout the stages of organic matter breakdown the color imparted to the soil varies from browns to black. Sodium content influences the depth of color of organic matter and therefore the soil. Sodium causes the organic matter (humus) to disperse more readily and spread over the soil particles, making the soil look darker (blacker). Soils which accumulate charcoal exhibit a black color.

Classification

Often described by using general terms, such as dark brown, yellowish brown, etc., soil colors are also described more technically by using Munsell soil color charts, which separate color into components of hue (relation to red, yellow and blue), value (lightness or darkness) and chroma (paleness or strength).

Types of Soil Color

Clear or White Soil Color

They may be white from which they may be influenced by calcium and magnesium carbonates, gypsum or other more soluble salts. Often, a white layer, mostly quartz occurs between organic matter on the surface where pigments were removed.

Brown Soil Color

Brown soils might be brown from decaying plant material. The darker color often indicates an increase in decomposed organic matter known as humus. Soil has living organisms and dead organic matter, which decomposes into black humus. In grassland (prairie) soils the dark color permeates through the surface layers bringing with it nutrients and high fertility Deeper in the soil, the organic pigment coats surfaces of soil, making them darker than the color inside. Humus color decreases with depth and iron pigments become more apparent. In forested areas, organic matter (leaves, needles, pine cones, dead animals) accumulates on top of the soil. Water-soluble carbon moves down through the soil and scavenges bits of humus and iron that accumulate below in black, humic bands over reddish iron bands.

Yellow or Red Soil Color

Yellow or red soil indicates the presence of oxidized ferric iron oxides. The red color might be mainly due to ferric oxides occurring as thin coatings on the soil particles while the iron oxide occurs as hematite or as hydrous ferric oxide, the color is red and when it occurs in the hydrate form as limonite the soil gets a yellow color.

Grey Soil Color

Organic matter plays an indirect, but crucial role in the removal of iron and manganese pigments in wet soils. All bacteria, including those that reduce iron and manganese, must have a food source. Therefore, anaerobic bacteria thrive in concentrations of organic matter, particularly in dead roots. Here, concentrations of gray mottles develop.

Factors affecting Soil Color

There are various factors or soil constituents that influence the soil color which are as follows:

- Organic Matter: Soils containing high amount of organic matter show the color variation from black to dark brown.

- Iron Compounds: Soils containing higher amount of iron compounds generally impart red, brown and yellow tinge color.

- Silica, Lime and Other Salts: Sometimes soils contain either large amounts of silica and lime or both. Due to presence of such materials in the soil the color of the soil appears like white or light colored.

- Mixture of Organic Matter and Iron Oxides: Very often soil contains a certain amount of organic matter and iron oxides. As a result of their existence in soil, the most common soil color is found and known as brown.

- Alternate Wetting and Drying Condition: During monsoon period due to heavy rain the reduction of soil occurs and during dry period the oxidation of soil also takes place. Due to development of such alternating oxidation and reduction condition, the color of soil in different horizons of the soil profile is variegated or mottled. This mottled color is due to residual products of this process especially iron and manganese compounds.

- Oxidation-Reduction Conditions: When soils are waterlogged for a longer period, the permanent reduced condition will develop. The presence of ferrous compounds resulting from the reducing condition in waterlogged soils impart bluish and greenish color.

Therefore, it may be concluded that the soil color indirectly indicative of many other important soil properties. Besides soil color directly modify the soil temperature e.g. dark colored soils absorb more heat than light colored soils.

Determination of Soil Color

The soil colors are best determined by the comparison with the Munsell color. This color chart is commonly used for this purpose. The color of the soil is a result of the light reflected from the soil.

Soil color rotation is divided into three parts:

- Hue: It denotes the dominant spectral color (red, yellow, blue and green).

- Value: It denotes the lightness or darkness of a color (the amount of reflected light).

- Chroma: It represents the purity of the color (strength of the color). The Munsell color notations are systematic numerical and letter designations of each of these three variables (hue, value and chroma). For example, the numerical notation 2.5 YR 5/6 suggests a hue of 2.5 YR, value of 5 and chroma of 6. The equivalent or parallel soil color name for this Munsell notation is 'red'.

Implication of Soil Color

- Color is one of the most useful and important characteristics for identification and classification of soils.

- Color often help to distinguish the different horizons of a soil profile, for example A_1 horizon is darker and B_2 horizon is brighter than adjacent horizons.

- Sometimes color helps in identifying diagnostic horizons used in soil classification such as a molicepipedon has a color so dark that it's both value and chroma are 3 or less.

- Soil color is an indicator of the soil moisture regimes under which a soil was developed. When a dry soil becomes moist, soil colors become darker by 1/2 to 3 steps in value, may change from − 1/2 to +2 steps in chroma and there is seldom change of hue.

Some of the largest difference in value between dry and moist colors occurs in grey or greyish brown horizons having moderate to moderately low contents of organic matter.

- As oxidation and reduction changes the color of iron containing minerals, the hydraulic regime or drainage status of a soil can be predicted from soil color. Bright (high chroma) colors such as red color throughout the profile are symptoms of well-drained soil through which water drains out easily and in which plenty of oxygen is available.

Prolonged water logging condition reduces iron oxide coatings and changes high chroma (red or brown) colors to low chroma (grey, bluish or grey-green) colors i.e. gray color.

Presence of gray (low chroma color) either alone or mixed in molted form with various shades of gray, brown and yellow is used in identifying imperfectly and poorly drained soils or is indicative of water logged conditions during at-least a major part of the plant growing season.

SOIL STRUCTURE

The arrangement of soil particles and their aggregate into certain defined patterns is called structure. The primary soil particles - sand, silt and clay usually occur grouped together in the form of aggregates. Natural aggregates are called peds, whereas clod is an artificially formed soil mass.

Structure is studied in the field under natural conditions and it is described under three categories:

- Type - Shape or form and arrangement pattern of peds.

- Class - Size of peds.

- Grade - Degree of distinctness of peds.

Types of Soil Structure

There are Four Principal Forms of Soil Structure

- Plate-like: In this structural type of aggregates are arranged in relatively thin horizontal plates. The horizontal dimensions are much more developed than the vertical. When the units are thick, they are called platy, and when thin, laminar. Platy structure is most noticeable in the surface layers of virgin soils but may be present in the sub-soil. Although most structural features are usually a product of soil forming forces, the platy type is often inherited from the parent material, especially those laid down by water.

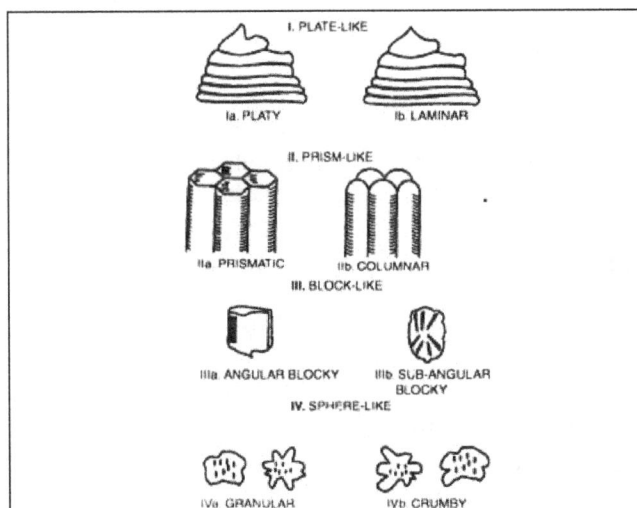

Types of Soil Structure.

- Prism-like: The vertical axis is more developed than horizontal, giving a pillar-like shape. When the top of such a ped is rounded, the structure is termed as columnar, and when flat, prismatic. They commonly occur in sub-soil horizons in arid and semi-arid regions.

- Block-like: All these dimensions are about the same size and the peds are cube-like with flat or rounded faces. When the faces are flat and the edges sharp angular, the structure is named as angular blocky. When the faces and edges are mainly rounded it is called sub angular blocky. These types usually are confined to the sub-soil and characteristics have much to do with soil drainage, aeration and root penetration.

- Spheroidal (Sphere-like): All rounded aggregates (peds) may be placed in this category, although the term more properly refers to those not over 0.5 inch in diameter. Those rounded complexes usually lie loosely and separately.

When wetted, the intervening spaces generally are not closed so readily by swelling as may be the case with a blocky structural condition. Therefore in sphere-like structure infiltration, percolation and aeration are not affected by wetting of soil. The aggregates of this group are usually termed as granular which are relatively less porous; when the granules are very porous, the term used is crumby.

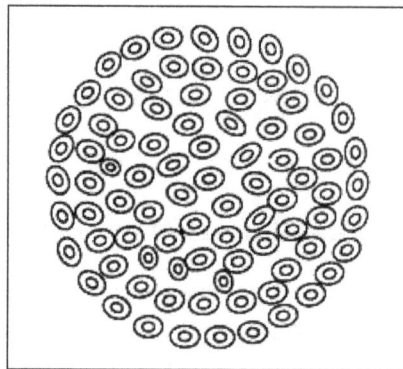

Sphere like soil structure.

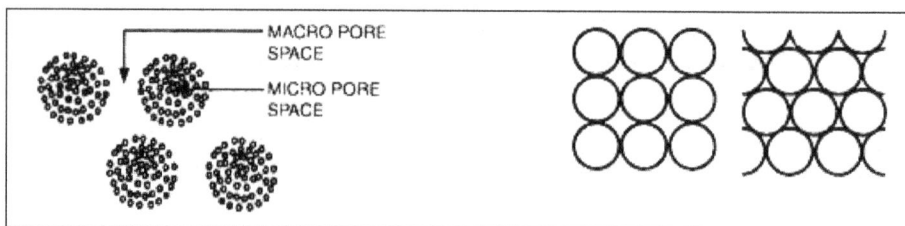

Arrangement of particles in Sphere-like soil structure and
Arrangment of Spherical particales in two ways.

Classes of Soil Structure

Each primary structural type of soil is differentiated into 5 size-classes depending upon the size of the individual peds.

The terms commonly used for the size classes are:

- Very fine or very thin,

- Fine or thin,

- Medium,

- Coarse or thick,

- Very coarse or very thick.

The terms thin and thick are used for platy types, while the terms fine and coarse are used for other structural types.

Grades of Soil Structure

Grades indicate the degree of distinctness of the individual peds. It is determined by the stability of the aggregates. Grade of structure is influenced by the moisture content of the soil. Grade also depends on organic matter, texture etc.

Four terms commonly used to describe the grade of soil structure are:

- Structure-less: There are no noticeable peds, such as conditions exhibited by loose sand or a cement-like condition of some clay soils.

- Weak structure: Indistinct formation of peds which are not durable.

- Moderate structure: Moderately well-developed peds which are fairly distinct.

- Strong structure: Very well-formed peds which are quite durable and distinct. For naming a soil structure the sequence followed is grade, class and type; for example, strong coarse angular blocky (soil structure).

Examples of Sphere-like Soil Structure

Often compound structures are met within the soil under natural conditions. For example, large prismatic types may break into medium blocky structure, constitute the compound structure.

	Grade	Class	Type	Structural name
1.	Strong	Fine	Granular	Strong fine granular
2.	Moderate	Very coarse	Granular	Moderate very coarse granular

Formation of Soil Structure

The mechanism of structure (aggregate) formation is quite complex. In aggregate formation a number of primary particles such as sand, silt and clay are brought together by the cementing or binding effect of soil colloidal clay, iron and aluminium hydroxides and organic matter.

The mineral Colloids (colloidal clay) by virtue of their properties of adhesion and cohesion, stick together to form aggregates. Sand and silt particles cannot form aggregates as they do not possess the power of adhesion and cohesion.

The amount and nature of colloidal clay influence the formation of aggregates. The greater the amount of clay in a soil, the greater is the tendency to form aggregates. Clay particles smaller than 0.001 mm aggregate very readily. So, clay minerals that have high base exchange capacity form aggregate more readily than those which have a low base exchange capacity. Iron and aluminium hydroxides act as cementing agent is binding the soil particles together. These are also responsible for forming aggregates by cementing sand and silt particles.

Organic matter plays an important part in forming soil aggregates. During decomposition of organic matter, humic acid and other sticky materials are produced which helps to form aggregate. Some fungi and bacteria taking part in the decomposition have also been found to have a cementing effect.

Another view of structure formation is that clay particles adsorbed by humus forming a clay-humus complex. It seems that humus absorbs both cations and anions. In normal soil, calcium is the predominant cation and forms calcium humate in combination with humus.

Factors affecting Soil Structure

The development of structure in arable soil depends on the following factors:

- Climate: Climate has considerable influence on the degree of aggregation as well as 011 the type of structure. In arid region, there is very little aggregation of primary particles. In semi- arid regions, the degree of aggregation is greater than arid regions.

- Organic matter: Organic matter improves the structure of a sandy soil as well as of a clay soil. In a case of sandy soil, the sticky and slimy material produced by the decomposing organic matter and the associated microorganism cement the sand particles to form aggregates. In the case of clayey soil, it modifies the properties of clay by reducing its cohesive power. This helps making clay more crumby.

- Tillage: Cultivation implements break down of large clods into smaller fragments and aggregates. For obtaining good granular and crumby structure, an optimum moisture content in the soil is necessary. If the moisture content is too high it will form large clods on drying. If it is too low, some of the existing aggregates will be broken down.

- Plant roots: Large number of granules remain attached to roots and root hairs which help to develop crumb structure. Plant root secretions may also act as cementing agents in binding the soil particles. The plant roots, on decay, may also bring about granulation due to the production of sticky substances.

- Soil organism: Among the soil fauna, small animals like earthworms, moles and insects etc., that burrow in the soil are the chief agents that take part in the aggregation of finer particles.

- Fertilizers: Fertilizer like Sodium nitrate destroys granulation by reducing the stability of aggregates. Few fertilizers, for example, Calcium Ammonium nitrate, help in development of good structures.

- Wetting and drying: Wren a dry soil is wetted, the soil colloids swell on absorbing water. On drying, shrinkage produced strains in the soil mass give rise to cracks which break it up into clods and granules of various sizes.

Effects of Soil Structure on other Physical Properties of Soil

Soil structure brings change in other physical properties of soil like porosity, temperature, density, consistency and color:

- Porosity: Porosity of a soil is easily changed. In plate-like structure pore spaces are less whereas in crumby structure pore spaces are more.

- Temperature: Crumby structure provides good aeration and percolation in the soil. Thus, these characteristics help in keeping optimum temperature in comparison to plate-like structure.

- Density: Bulk density varies with the total pore space present in the soil. Structure chiefly influences pore spaces. Platy structure with less total pore spaces has high bulk density whereas crumby structure with more total pore spaces has low bulk density.

- Consistence: Consistence of soil also depends on structure. Plate-like structure exhibits strong plasticity.

- Color: Bluish and greenish colors of soil are generally due to poor drainage of soil. Platy structure normally hinders free drainage.

Structural Management of Soils

- Coarse-textured Soil: Sandy soils are commonly too loose and lack the capacity to adsorb and hold sufficient moisture and nutrients. They lack fertility and water- holding capacity. There is only one practical method of improving the structure of such soil- the addition of organic matter. Organic matter will not only act as a binding agent for the particles but will also increase the water-holding capacity. Sod-crops, for example, corn, blue grass etc., also help in improving the structural condition of sandy soils.

- Fine-textured Soil: The structural management of a clay soil is difficult than sandy soil. In clay, plasticity and cohesion are high because of the presence of large amount of colloidal clay. When such a soil is tilled when wet, its pore space becomes much reduced, it becomes practically impervious to air and water and it is said to be puddled. When a soil in this condition dries, it usually becomes hard and dense. The tillage of clay soil should be done at right moisture stage. If ploughed too wet, the structural aggregates are broken down and an un-favourable structure results. On the other hand, if ploughed too dry, big clods are turned up which are difficult to work. The granulation of fine-textured soil should be encouraged by the incorporation of organic matter. Growing of sod-crops also improves granulation in the soil.

- Rice Soil: Puddling of the soil is generally beneficial to the production of rice. In preparation for the planting of rice, the soil is flooded with water and then puddled by intensive tillage. Puddling destroys the structural aggregates. Rice seedling is transplanted into the freshly prepared mud.

Such soil management helps control weeds and also reduce the rate of water movement down (percolation) through the soil. This is important to maintain standing water in the rice through out the growing season. By reducing water percolation, puddled soil markedly decreases the amount of water needed to produce a rice crop.

Semi-aquatic characteristics of the rice plant account for its positive response to a type of soil management that destroy aggregate. Rice survives flooded conditions because oxygen moves downward inside the stem of the plant to supply the roots. This characteristic permits rice to stand well in the water-logging condition. Rice can be grown successfully on un-puddled but flooded soil.

PORE SPACE IN SOIL

The pore space of soil contains the liquid and gas phases of soil, i.e., everything but the solid phase that contains mainly minerals of varying sizes as well as organic compounds.

In order to understand porosity better a series of equations have been used to express the quantitative interactions between the three phases of soil.

Macropores or fractures play a major role in infiltration rates in many soils as well as preferential flow patterns, hydraulic conductivity and evapotranspiration. Cracks are also very influential in gas exchange, influencing respiration within soils. Modeling cracks therefore helps understand how these processes work and what the effects of changes in soil cracking such as compaction, can have on these processes.

Bulk Density

$$\rho = \frac{M_s}{V_t}.$$

The bulk density of soil depends greatly on the mineral make up of soil and the degree of compaction. The density of quartz is around 2.65 g/cm³ but the bulk density of a soil may be less than half that density.

Most soils have a bulk density between 1.0 and 1.6 g/cm³ but organic soil and some friable clay may have a bulk density well below 1 g/cm³.

Core samples are taken by driving a metal core into the earth at the desired depth and soil horizon. The samples are then oven dried and weighed.

Bulk density = (mass of oven dry soil)/volume

The bulk density of soil is inversely related to the porosity of the same soil. The more pore space in a soil the lower the value for bulk density.

Porosity

$$f = \frac{V_f}{V_t} \quad \text{or} f = \frac{V_a + V_w}{V_s + V_a + V_w}$$

Porosity is a measure of the total pore space in the soil. This is measured as a volume or percent. The amount of porosity in a soil depends on the minerals that make up the soil and the amount of sorting that occurs within the soil structure. For example, a sandy soil will have larger porosity than silty sand, because the silt will fill in the gaps between the sand particles.

Pore Space Relations

Hydraulic Conductivity

Hydraulic conductivity (K) is a property of soil that describes the ease with which water can move through pore spaces. It depends on the permeability of the material (pores, compaction) and on the degree of saturation. Saturated hydraulic conductivity, K_{sat}, describes water movement through saturated media. Where hydraulic conductivity has the capability to be measured at any state. It can be estimated by numerous kinds of equipment. To calculate hydraulic conductivity, Darcy's law is used. The manipulation of the law depends on the Soil saturation and instrument used.

Infiltration

Infiltration is the process by which water on the ground surface enters the soil. The water enters the soil through the pores by the forces of gravity and capillary action. The largest cracks and pores offer a great reservoir for the initial flush of water. This allows a rapid infiltration. The smaller pores take longer to fill and rely on capillary forces as well as gravity. The smaller pores have a slower infiltration as the soil becomes more saturated.

Pore Types

A pore is not simply a void in the solid structure of soil. The various pore size categories have different characteristics and contribute different attributes to soils depending on the number and frequency of each type. A widely used classification of pore size is that of Brewer:

- Macropore: The pores that are too large to have any significant capillary force. Unless impeded, water will drain from these pores, and they are generally air-filled at field capacity. Macropores can be caused by cracking, division of peds and aggregates, as well as, plant roots, and zoological exploration. Size >75 µm.

- Mesopore: The largest pores filled with water at field capacity. Also known as storage pores because of the ability to store water useful to plants. They do not have capillary forces too great so that the water does not become limiting to the plants. The properties of mesopores are highly studied by soil scientists because of their impact on agriculture and irrigation. Size 30 µm–75 µm.

- Micropore: These are "pores that are sufficiently small that water within these pores is considered immobile, but available for plant extraction."Because there is little movement of water in these pores, solute movement is mainly by the process of diffusion. Size 5-30 µm.

- Ultramicropore: These pores are suitable for habitation by microorganisms. Their distribution is determined by soil texture and soil organic matter, and they are not greatly affected by compaction. Size 0.1-30 µm.

- Cryptopore: Pores that are too small to be penetrated by most microorganisms. Organic matter in these pores is therefore protected from microbial decomposition. They are filled with water unless the soil is very dry, but little of this water is available to plants, and water movement is very slow. Size <0.1 μm.

Modelling Methods

Basic crack modeling has been undertaken for many years by simple observations and measurements of crack size, distribution, continuity and depth. These observations have either been surface observation or done on profiles in pits. Hand tracing and measurement of crack patterns on paper was one method used prior to advances in modern technology. Another field method was with the use of string and a semicircle of wire. The semi circle was moved along alternating sides of a string line. The cracks within the semicircle were measured for width, length and depth using a ruler. The crack distribution was calculated using the principle of Buffon's needle.

Disc Permeameter

This method relies on the fact that crack sizes have a range of different water potentials. At zero water potential at the soil surface an estimate of saturated hydraulic conductivity is produced, with all pores filled with water. As the potential is decreased progressively larger cracks drain. By measuring at the hydraulic conductivity at a range of negative potentials, the pore size distribution can be determined. While this is not a physical model of the cracks, it does give an indication to the sizes of pores within the soil.

Horgan and Young Model

Horgan and Young produced a computer model to create a two-dimensional prediction of surface crack formation. It used the fact that once cracks come within a certain distance of one another they tend to be attracted to each other. Cracks also tend to turn within a particular range of angles and at some stage a surface aggregate gets to a size that no more cracking will occur. These are often characteristic of a soil and can therefore be measured in the field and used in the model. However it was not able to predict the points at which cracking starts and although random in the formation of crack pattern, in many ways, cracking of soil is often not random, but follows lines of weaknesses.

Araldite-impregnation Imaging

A large core sample is collected. This is then impregnated with araldite and a fluorescent resin. The core is then cut back using a grinding implement, very gradually and at every interval the surface of the core sample is digitally imaged. The images are then loaded into a computer where they can be analysed. Depth, continuity, surface area and a number of other measurements can then be made on the cracks within the soil.

Electrical Resistivity Imaging

Using the infinite resistivity of air, the air spaces within a soil can be mapped. A specially designed resistivity meter had improved the meter-soil contact and therefore the area of the reading. This technology can be used to produce images that can be analysed for a range of cracking properties.

SOIL THERMAL PROPERTIES

The thermal properties of soil are a component of soil physics that has found important uses in engineering, climatology and agriculture. These properties influence how energy is partitioned in the soil profile. While related to soil temperature, it is more accurately associated with the transfer of energy (mostly in the form of heat) throughout the soil, by radiation, conduction and convection.

The main soil thermal properties are:

- Volumetric heat capacity, SI Units: $J \cdot m^{-3} \cdot K^{-1}$.

- Thermal conductivity, SI Units: $W \cdot m^{-1} \cdot K^{-1}$.

- Thermal diffusivity, SI Units: $m^2 \cdot s^{-1}$.

Measurement

It is hard to say something general about the soil thermal properties at a certain location because these are in a constant state of flux from diurnal and seasonal variations. Apart from the basic soil composition, which is constant at one location, soil thermal properties are strongly influenced by the soil volumetric water content, volume fraction of solids and volume fraction of air. Air is a poor thermal conductor and reduces the effectiveness of the solid and liquid phases to conduct heat. While the solid phase has the highest conductivity it is the variability of soil moisture that largely determines thermal conductivity. As such soil moisture properties and soil thermal properties are very closely linked and are often measured and reported together. Temperature variations are most extreme at the surface of the soil and these variations are transferred to sub surface layers but at reduced rates as depth increases. Additionally there is a time delay as to when maximum and minimum temperatures are achieved at increasing soil depth (sometimes referred to as thermal lag).

One possible way of assessing soil thermal properties is the analysis of soil temperature variations versus depth Fourier's law,

$$Q = -\lambda dT / dz$$

where Q is heat flux or rate of heat transfer per unit area $J \cdot m^{-2} \cdot s^{-1}$ or $W \cdot m^{-2}$, λ is thermal conductivity $W \cdot m^{-1} \cdot K^{-1}$; dT/dz is the gradient of temperature (change in temp/change in depth) $K \cdot m^{-1}$.

The most commonly applied method for measurement of soil thermal properties, is to perform in-situ measurements, using Non-Steady-State Probe systems or Heat Probes.

Single and Dual Heat Probes

The single probe method employs a heat source inserted into the soil whereby heat energy is applied continuously at a given rate. The thermal properties of the soil can be determined by analysing the temperature response adjacent to the heat source via a thermal sensor. This method reflects the rate at which heat is conducted away from the probe. The limitation of this device is that it measures thermal conductivity only. Applicable standards are: IEEE Guide for Soil Thermal Resistivity Measurements as well as with ASTM D 5334-08 Standard Test Method for Determination of Thermal Conductivity of Soil and Soft Rock by Thermal Needle Probe Procedure.

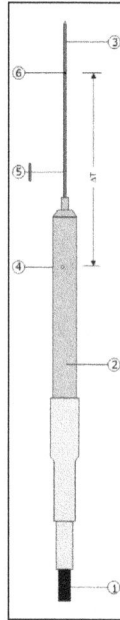

Small Size Non-Steady-State Probe: The probe consists of a needle (3) with a single thermocouple junction (6) and a heating wire, (5). It is inserted into the medium that is investigated.

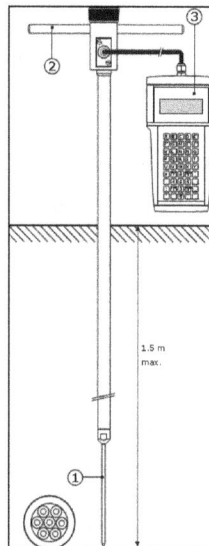

A complete system for measurement of soil thermal conductivity, specially designed for measurements at around 1.5 meters below the soil surface, which is the typical depth of burial for high voltage cables.

After further research the dual-probe heat-pulse technique was developed. It consists of two parallel needle probes separated by a distance. One probe contains a heater and the other a temperature sensor. The dual probe device is inserted into the soil and a heat pulse is applied and the temperature sensor records the response as a function of time. That is, a heat pulse is sent from the probe across the soil to the sensor. The great benefit of this device is that it measures both thermal diffusivity and volumetric heat capacity. From this, thermal conductivity can be calculated meaning the dual probe can determine all the main soil thermal properties. Potential drawbacks of the heat-pulse technique have been noted. This includes the small measuring volume of soil as well as measurements being sensitive to probe-to-soil contact and sensor-to-heater spacing.

Remote Sensing

Remote sensing from satellites, aircraft has greatly enhanced how the variation in soil thermal properties can be identified and utilized to benefit many aspects of human endeavor. While remote sensing of reflected light from surfaces does indicate thermal response of the topmost layers of soil (a few molecular layers thick), it is thermal infrared wavelength that provides energy variations extending to varying shallow depths below the ground surface which is of most interest. A thermal sensor can detect variations to heat transfers into and out of near surface layers because of external heating by the thermal processes of conduction, convection, and radiation. Microwave remote sensing from satellites has also proven useful as it has an advantage over TIR of not being effected by cloud cover.

The various methods of measuring soil thermal properties have been utilized to assist in diverse fields such as; the expansion and contraction of construction materials especially in freezing soils, longevity and efficiency of gas pipes or electrical cables buried in the ground, energy conservation schemes, in agriculture for timing of planting to ensure optimum seedling emergence and crop growth, measuring greenhouse gas emissions as heat effects the liberation of carbon dioxide from soil. Soil thermal properties are also becoming important in areas of environmental science such as determining water movement in radioactive waste and in locating buried land mines.

Uses

The thermal inertia of the soil enables the ground to be used for underground thermal energy storage. Solar energy can be recycled from summer to winter by using the ground as a long term store of heat energy before being retrieved by ground source heat pumps in winter.

Changes in the amount of dissolved organic carbon and soil organic carbon within soil can effect its ability to respirate, either increasing or decreasing the soils carbon uptake.

Furthermore, MCS design criteria for shallow loop ground source heat pumps require an accurate in situ thermal conductivity reading. This can be done by using the above-mentioned thermal heat probe to accurately determine soil thermal conductivity across the site.

SOIL RESISTIVITY

The measure of the resistance offered by the soil in the flow of electricity, is called the soil resistivity. The resistivity of the soil depends on the various factors likes soil composition, moisture, temperature, etc. Generally, the soil is not homogenous, and their resistivity varies with the depth.

The soil having a low resistivity is good for designing the grounding system. The resistivity of the soil is measured in ohmmeter or ohm-centimeters.

The resistivity of the soil mainly depends on its temperature. When the temperature of the soil is more than 0°, then its effect on soil resistivity is negligible. At 0° the water starts freezing and resistivity increases. The magnitude of the current also affects the resistivity of the soil. If the magnitude of current dissipated in the soil is high, it may cause significant drying of soil and increase its resistivity.

The resistivity of the soil varies with the depth. The lower layers of the soil have greater moisture content and lower resistivity. If the lower layer contains hard and rocky layers, then their resistivity may increase with the depth.

Measurement of Soil Resistivity

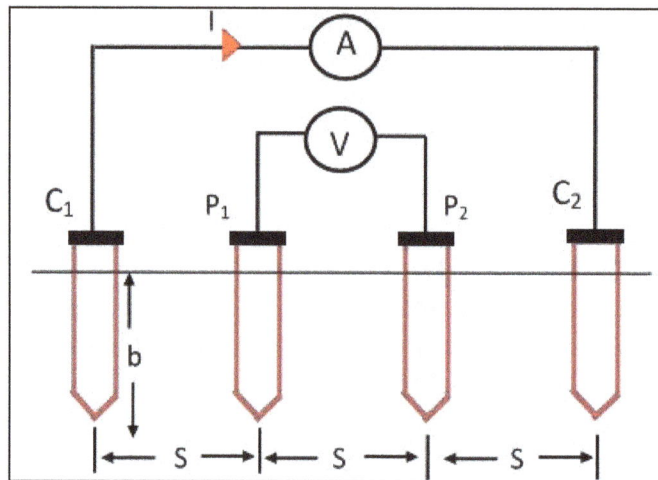

Measurement of Soil Resistivity.

The resistivity of the soil is usually measured by the four spike methods. In this method the four spikes arranged in the straight line are driven into the soil at equal distance. A known current is passed between electrode C_1 and C_2 and potential drop V is measured across P_1 and P_2. The current I developed an electric field which is proportional to current density and soil resistivity. The voltage V is proportional to this field.

The soil resistivity is proportional to the ratio of the voltage V and current I and is given as:

$$\rho = \frac{\dfrac{4\pi SV}{I}}{1 + \dfrac{2S}{\sqrt{S^2 + 4b^2}} - \dfrac{2S}{\sqrt{4S^2 + 4b^2}}}.$$

Where ρ is the resistivity of the soil and their unit is ohmmeters. S is the horizontal space between the spikes in m and b is the depth of burial in metre.

If the measurement is to be carried out using the main supply, an isolating transformer should be connected between the main supply and the test setup. So that the result may not be affected by it.

References

- Soilphysics: psc.usu.edu, Retrieved 11 June, 2019
- Vrieze, Jop de (2015-08-14). "The littlest farmhands". Science. 349 (6249): 680–683. Doi:10.1126/science.349.6249.680. ISSN 0036-8075. PMID 26273035
- Physics: soils4teachers.org, Retrieved 12 July, 2019
- Soilsmatter2011 (2015-06-30). "What types of animals live in the soil? Why is soil condition important to them?". Soils Matter, Get the Scoop!. Retrieved 2019-05-08
- Chemistry: soils4teachers.org, Retrieved 13 January, 2019
- Saltini Antonio, Storia delle scienze agrarie, 4 vols, Bologna 1984-89, ISBN 88-206-2412- 5, ISBN 88-206-2413-3, ISBN 88-206-2414-1, ISBN 88-206-2415-X
- Bockheim, J.G.; Gennadiyev, A.N.; Hammer, R.D.; Tandarich, J.P. (January 2005). "His-torical development of key concepts in pedology" (PDF). Geoderma. 124 (1–2): 23–36. Doi:10.1016/j.geoderma.2004.03.004. Retrieved 2 March 2016

Soil Quality

The ability of a soil to perform its functions which are necessary for environment and people is referred to as soil quality. Some of the aspects which are studied under the domain of soil quality include soil pH, soil acidification, soil organic matter and soil salinity. This chapter closely examines these aspects of soil quality to provide an extensive understanding of the subject.

Soil quality is the capacity of a soil to function for specific land uses or within ecosystem boundaries. This capacity is an inherent characteristic of a soil and varies from soil to soil. Such indicators as organic-matter content, salinity, tilth, compaction, available nutrients, and rooting depth help measure the health or condition of the soil-its quality-in any given place.

For example, organic-matter content, biological activity, acidity, and salinity are related to the ability of a soil to store and cycle nutrients for plant growth. Soil tilth, compaction, and available water capacity reflect the ability of a soil to regulate and partition the flow of water. Texture, such as loam or clay, is an important soil property in the support of buildings and roads. An enhancement of soil health or quality could be measured by an increase in organic matter content in cultivated soils over the years, which would reflect the soil's ability to cycle nutrients.

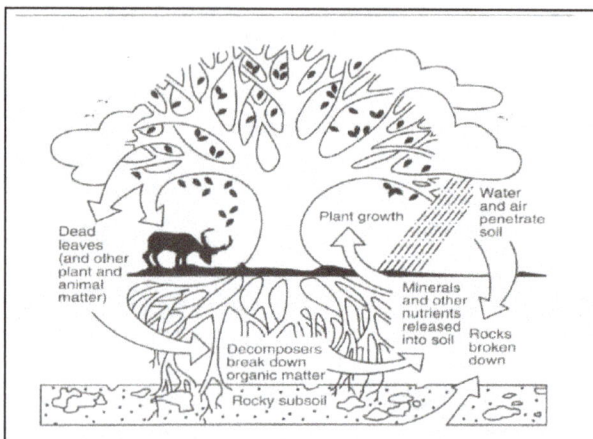

If, through human use, the ability of a soil to perform certain functions is maintained or enhanced, its quality contributes to sustain-ability. However, if the ability of a soil to perform beneficial functions has been impaired, the quality or condition of that soil has been degraded and that land use detracts from the sustainability of the soil resource.

Factors Reducing Soil Quality

Every year in every country, soil resources are impaired and in some cases lost for productive use because of misuse, application of toxic materials, or poor land management systems. This happens because:

- Lack of knowledge about individual soils and their properties as they relate to soil quality and land use.

- Lack of understanding of the impact of management activities on individual soils and the ability of the soil to maintain beneficial functions.

- Excessive human demands on available soil resources, thus forcing use of more fragile or erodible lands for production of food and fiber.

A major cause of reduced soil quality is soil erosion, the removal of the topsoil. Although soil erosion is a natural geologic process, it is often accelerated by cultivation and resource development to meet human needs. However, at least 40 percent of United States erosion losses result from nonagricultural activities and events, such as logging, construction, off-road vehicles, floods, droughts, and fires. Erosion degrades soil condition by lowering organic-matter content, decreasing rooting depth, and decreasing available water capacity. In addition, sediment (the product of erosion) can pollute streams, lakes, and other bodies of water with soil particles and associated chemicals and plant nutrients. Soil erosion by wind pollutes the air and can damage plants through a sandblasting effect.

Compaction, accumulation of salts, excess nutrients and chemicals, and toxic chemicals are also significant soil quality concerns. Over application of plant nutrients and pesticides can exceed the capacity of a soil to cycle nutrients efficiently. This increases the potential for those nutrients and chemicals to leach from a soil in percolating waters and eventually pollute ground and surface waters. Some crop management practices, such as excessive tillage, compact the soil, hampering the ability of the soil to receive rainwater; thus, more water runs off the soil surface instead of entering it and being stored for plant use. Although natural soil processes can transform substances such as sewage sludge and animal wastes into nutrients for plant growth, over application of these materials will impair the ability of a soil to transform them-again, leading to a hazard of ground or surface water pollution.

Improving Soil Quality

The sodbuster and conservation compliance provisions in the Food Security Act of 1985 have led to increased application of conservation cropping systems, which-along with on-going conservation efforts on America's farms and ranches and application of crop residue management technologies-have significantly cut soil erosion. Sheet & rill erosion (the removal of soil in thin layers sheets or tiny channels rills) and wind erosion on the Nation's cropland declined from 3.1 billion tons in 1982 to 2.1 billion tons in 1992. In addition, these conservation systems have increased the organic-matter content of soils and improved the tilth of their surface layers.

The Natural Resources Conservation Service has adopted a holistic approach to help land managers protect soil and related resources. This approach assesses the natural processes and functions of an area, such as a watershed, and determines how these functions affect the quality of the soil, water, air, plant, and animal resources. The holistic approach incorporates land managers' objectives and resource concerns and then provides a set of conservation practices that will maintain or enhance soil and environmental quality.

Soil scientists and others are developing tools to measure and monitor soil quality by developing key indicators to assess soil condition or health. While monitoring is important for assessing change, it alone will not improve soil quality. Monitoring does help provide feedback on how prescribed conservation management systems are performing and provides conservationists with data to design needed modifications in these systems.

Dominant land uses in the United States. People generally put their soils to uses that the soils can support; in this way land use reflects soil quality.

Approaches to Maintaining Soil Quality

For the long term, the best way to maintain or improve soil quality is to locate agricultural production on the soils that are best suited for the particular use, and use other soils according to their capacity. Beyond that, we should remediate soil quality problems and address areas of greatest concern not only for the soils, but also for related water and air resources. We should also develop technologies that maximize the ability of soils to function for specific land uses-such as soil-specific nutrient and pesticide management, no-till and other crop residue management technologies, appropriate use of urban and animal wastes, and conservation management systems. And finally, we should develop the public's awareness and concern for the quality of the soil resource and how it affects their daily lives.

Benefits of Soil Quality

High-quality soils ensure that the primary agricultural lands in the United States are sustained for future generations. Soils of high quality are essential for the production of a bountiful supply of safe food and fiber. Healthy food translates to a healthy people and a healthy nation. High-quality soils support:

- Clean water by transforming harmful substances and chemicals to nontoxic forms, cycling nutrients, and partitioning rainfall to keep sediments and chemicals out of lakes and streams.

- Clean and healthy air by keeping dust particles out of the air and cycling other gases.

- Healthy plant growth by storing nutrients and water and providing structural support through a receptive rooting medium.

- Storage of greenhouse gases such as carbon dioxide in the form of organic matter in the soil.

The primary benefit of enhanced soil quality is the protection of a finite resource. Maintenance and enhancement of soil quality maintains maximum efficiency in crop productivity over time by enhancing nutrient cycling and encouraging site-specific application of nutrients and pesticides. It protects water and air quality and preserves the beneficial functions of the soil in specific ecosystems.

Soil and soil quality are closely linked with the other natural resources-water, air, plants, and animals. As soil quality improves, so does the quality of the other resources. The key is using management practices to care for the land in a way that improves soil quality-to better filter water, reduce airborne particles, maintain or improve productivity, use less chemicals, and increase plant, animal and micro-organism diversity.

SOIL pH

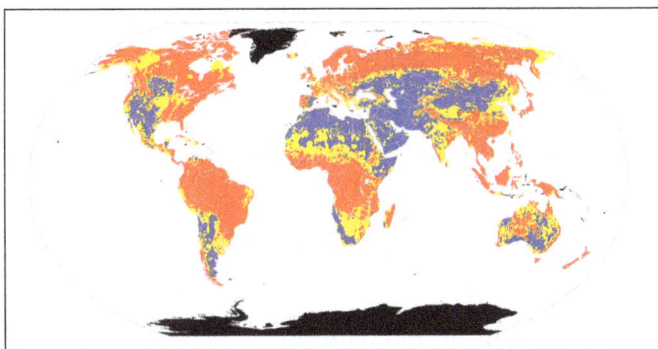

Global variation in soil pH. Red = acidic soil. Yellow = neutral soil. Blue = alkaline soil. Black = no data.

Soil pH is a measure of the acidity or basicity (alkalinity) of a soil. pH is defined as the negative logarithm (base 10) of the activity of hydronium ions (H^+ or, more precisely, $H_3O^+_{aq}$) in a solution. In soils, it is measured in a slurry of soil mixed with water (or a salt solution, such as 0.01 M $CaCl_2$), and normally falls between 3 and 10, with 7 being neutral. Acid soils have a pH below 7 and alkaline soils have a pH above 7. Ultra-acidic soils (pH < 3.5) and very strongly alkaline soils (pH > 9) are rare.

Soil pH is considered a master variable in soils as it affects many chemical processes. It specifically affects plant nutrient availability by controlling the chemical forms of the different nutrients and influencing the chemical reactions they undergo. The optimum pH range for most plants is between 5.5 and 7.5; however, many plants have adapted to thrive at pH values outside this range.

Classification of Soil pH Ranges

The United States Department of Agriculture Natural Resources Conservation Service classifies soil pH ranges as follows:

Denomination	pH range
Ultra acidic	< 3.5
Extremely acidic	3.5–4.4

Very strongly acidic	4.5–5.0
Strongly acidic	5.1–5.5
Moderately acidic	5.6–6.0
Slightly acidic	6.1–6.5
Neutral	6.6–7.3
Slightly alkaline	7.4–7.8
Moderately alkaline	7.9–8.4
Strongly alkaline	8.5–9.0
Very strongly alkaline	> 9.0

Determining pH

Methods of determining pH include:

- Observation of soil profile: Certain profile characteristics can be indicators of either acid, saline, or sodic conditions.

  ○ Poor incorporation of the organic surface layer with the underlying mineral layer – this can indicate strongly acidic soils;

  ○ The classic podzol horizon sequence, since podzols are strongly acidic: in these soils, a pale eluvial (E) horizon lies under the organic surface layer and overlies a dark B horizon;

  ○ Presence of a caliche layer indicates the presence of calcium carbonates, which are present in alkaline conditions;

  ○ Columnar structure can be an indicator of sodic condition.

- Observation of predominant flora: Calcifuge plants (those that prefer an acidic soil) include *Erica, Rhododendron* and nearly all other Ericaceaespecies, many birch (*Betula*), foxglove (*Digitalis*), gorse (*Ulex* spp.), and Scots Pine (*Pinus sylvestris*). Calcicole (lime loving) plants include ash trees (*Fraxinus* spp.), honeysuckle (*Lonicera*), *Buddleja*, dogwoods (*Cornus* spp.), lilac (*Syringa*) and *Clematis* species.

- Use of an inexpensive pH testing kit, where in a small sample of soil is mixed with indicator solution which changes colour according to the acidity.

- Use of litmus paper: A small sample of soil is mixed with distilled water, into which a strip of litmus paper is inserted. If the soil is acidic the paper turns red, if basic, blue.

- Use of a commercially available electronic pH meter, in which a glass or solid-state electrode is inserted into moistened soil or a mixture (suspension) of soil and water; the pH is usually read on a digital display screen.

- Recently, spectrophotometric methods have been developed to measure soil pH involving addition of an indicator dye to the soil extract. These compared well to glass electrode measurements but offer substantial advantages such as lack of drift, liquid junction and suspension effects.

Precise, repeatable measures of soil pH are required for scientific research and monitoring. This generally entails laboratory analysis using a standard protocol.

Factors affecting Soil pH

The pH of a natural soil depends on the mineral composition of the parent material of the soil, and the weathering reactions undergone by that parent material. In warm, humid environments, soil acidification occurs over time as the products of weathering are leached by water moving laterally or downwards through the soil. In dry climates, however, soil weathering and leaching are less intense and soil pH is often neutral or alkaline.

Sources of Acidity

Many processes contribute to soil acidification. These include:

- Rainfall: Acid soils are most often found in areas of high rainfall. Rainwater has a slightly acidic pH (usually about 5.7) due to a reaction with CO_2 in the atmosphere that forms carbonic acid. When this water flows through soil it results in the leaching of basic cations from the soil as bicarbonates; this increases the percentage of Al^{3+} and H^+ relative to other cations.

- Root respiration and decomposition of organic matter by microorganisms releases CO_2 which increases the carbonic acid (H_2CO_3) concentration and subsequent leaching.

- Plant growth: Plants take up nutrients in the form of ions (e.g. NO^-_3, NH^+_4, Ca^{2+}, $H_2PO^-_4$), and they often take up more cations than anions. However plants must maintain a neutral charge in their roots. In order to compensate for the extra positive charge, they will release H^+ ions from the root. Some plants also exude organic acids into the soil to acidify the zone around their roots to help solubilize metal nutrients that are insoluble at neutral pH, such as iron (Fe).

- Fertilizer use: Ammonium (NH^+_4) fertilizers react in the soil by the process of nitrification to form nitrate (NO^-_3), and in the process release H^+ ions.

- Acid rain: The burning of fossil fuels releases oxides of sulfur and nitrogen into the atmosphere. These react with water in the atmosphere to form sulfuric and nitric acid in rain.

- Oxidative weathering: Oxidation of some primary minerals, especially sulfides and those containing Fe^{2+}, generate acidity. This process is often accelerated by human activity:

  - Mine spoil: Severely acidic conditions can form in soils near some mine spoils due to the oxidation of pyrite.

  - Acid sulfate soils formed naturally in waterlogged coastal and estuarine environments can become highly acidic when drained or excavated.

Sources of Alkalinity

Total soil alkalinity increases with:

- Weathering of silicate, aluminosilicate and carbonate minerals containing Na^+, Ca^{2+}, Mg^{2+} and K^+;

- Addition of silicate, aluminosilicate and carbonate minerals to soils; this may happen by deposition of material eroded elsewhere by wind or water, or by mixing of the soil with less weathered material (such as the addition of limestone to acid soils);

- Addition of water containing dissolved bicarbonates (as occurs when irrigating with high-bicarbonate waters).

The accumulation of alkalinity in a soil (as carbonates and bicarbonates of Na, K, Ca and Mg) occurs when there is insufficient water flowing through the soils to leach soluble salts. This may be due to arid conditions, or poor internal soil drainage; in these situations most of the water that enters the soil is transpired (taken up by plants) or evaporates, rather than flowing through the soil.

The soil pH usually increases when the total alkalinity increases, but the balance of the added cations also has a marked effect on the soil pH. For example, increasing the amount of sodium in an alkaline soil tends to induce dissolution of calcium carbonate, which increases the pH. Calcareoussoils may vary in pH from 7.0 to 9.5, depending on the degree to which Ca^{2+} or Na^+ dominate the soluble cations.

Effect of Soil pH on Plant Growth

Acid Soils

Plants grown in acid soils can experience a variety of stresses including aluminium (Al), hydrogen (H), and/or manganese (Mn) toxicity, as well as nutrient deficiencies of calcium (Ca) and magnesium (Mg).

Aluminium toxicity is the most widespread problem in acid soils. Aluminium is present in all soils, but dissolved Al^{3+} is toxic to plants; Al^{3+} is most soluble at low pH; above pH 5.0, there is little Al in soluble form in most soils. Aluminium is not a plant nutrient, and as such, is not actively taken up by the plants, but enters plant roots passively through osmosis. Aluminium inhibits root growth; lateral roots and root tips become thickened and roots lack fine branching; root tips may turn brown. In the root, the initial effect of Al^{3+} is the inhibition of the expansion of the cells of the rhizodermis, leading to their rupture; thereafter it is known to interfere with many physiological processes including the uptake and transport of calcium and other essential nutrients, cell division, cell wall formation, and enzyme activity.

Proton (H^+ ion) stress can also limit plant growth. The proton pump, H^+-ATPase, of the plasmalemma of root cells works to maintain the near-neutral pH of their cytoplasm. A high proton activity (pH within the range 3.0–4.0 for most plant species) in the external growth medium overcomes the capacity of the cell to maintain the cytoplasmic pH and growth shuts down.

In soils with a high content of manganese-containing minerals, Mn toxicity can become a problem at pH 5.6 and lower. Manganese, like aluminium, becomes increasingly soluble as pH drops, and Mn toxicity symptoms can be seen at pH levels below 5.6. Manganese is an essential plant nutrient, so plants transport Mn into leaves. Classic symptoms of Mn toxicity are crinkling or cupping of leaves.

Nutrient Availability in Relation to Soil pH

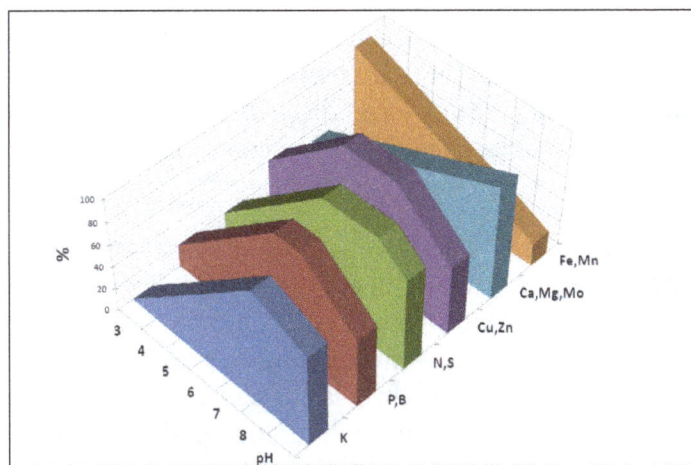

Nutrient availability in relation to soil pH.

Soil pH affects the availability of some plant nutrients:

- Aluminium toxicity has direct effects on plant growth; however, by limiting root growth, it also reduces the availability of plant nutrients. Because roots are damaged, nutrient uptake is reduced, and deficiencies of the macronutrients (nitrogen, phosphorus, potassium, calcium and magnesium) are frequently encountered in very strongly acidic to ultra-acidic soils (pH<5.0).

- Molybdenum availability is increased at higher pH; this is because the molybdate ion is more strongly sorbed by clay particles at lower pH.

- Zinc, iron, copper and manganese show decreased availability at higher pH (increased sorbtion at higher pH).

- The effect of pH on phosphorus availability varies considerably, depending on soil conditions and the crop in question. The prevailing view in the 1940s and 1950s was that P availability was maximized near neutrality (soil pH 6.5–7.5), and decreased at higher and lower pH. Interactions of phosphorus with pH in the moderately to slightly acidic range (pH 5.5–6.5) are, however, far more complex than is suggested by this view. Laboratory tests, glasshouse trials and field trials have indicated that increases in pH within this range may increase, decrease, or have no effect on P availability to plants.

Water Availability in Relation to Soil pH

Strongly alkaline soils are sodic and dispersive, with slow infiltration, low hydraulic conductivity and poor available water capacity. Plant growth is severely restricted because aeration is poor when the soil is wet; in dry conditions, plant-available water is rapidly depleted and the soils become hard and cloddy (high soil strength).

Many strongly acidic soils, on the other hand, have strong aggregation, good internal drainage, and good water-holding characteristics. However, for many plant species, aluminium toxicity severely limits root growth, and moisture stress can occur even when the soil is relatively moist.

Plant pH Preferences

In general terms, different plant species are adapted to soils of different pH ranges. For many species, the suitable soil pH range is fairly well known. Online databases of plant characteristics, such *USDA PLANTS* and *Plants for a Future* can be used to look up the suitable soil pH range of a wide range of plants. Documents like *Ellenberg's indicator values for British plants* can also be consulted.

However, a plant may be intolerant of a particular pH in some soils as a result of a particular mechanism, and that mechanism may not apply in other soils. For example, a soil low in molybdenum may not be suitable for soybean plants at pH 5.5, but soils with sufficient molybdenum allow optimal growth at that pH. Similarly, some calcifuges (plants intolerant of high-pH soils) can tolerate calcareous soils if sufficient phosphorus is supplied. Another confounding factor is that different varieties of the same species often have different suitable soil pH ranges. Plant breeders can use this to breed varieties that can tolerate conditions that are otherwise considered unsuitable for that species – examples are projects to breed aluminium-tolerant and manganese-tolerant varieties of cereal crops for food production in strongly acidic soils.

Changing Soil pH

Increasing pH of Acidic Soil

Finely ground agricultural lime is often applied to acid soils to increase soil pH (liming). The amount of limestone or chalk needed to change pH is determined by the mesh size of the lime (how finely it is ground) and the buffering capacity of the soil. A high mesh size (60 mesh = 0.25 mm; 100 mesh = 0.149 mm) indicates a finely ground lime that will react quickly with soil acidity. The buffering capacity of a soil depends on the clay content of the soil, the type of clay, and the amount of organic matter present, and may be related to the soil cation exchange capacity. Soils with high clay content will have a higher buffering capacity than soils with little clay, and soils with high organic matter will have a higher buffering capacity than those with low organic matter. Soils with higher buffering capacity require a greater amount of lime to achieve an equivalent change in pH.

Amendments other than agricultural lime that can be used to increase the pH of soil include wood ash, industrial calcium oxide (burnt lime), magnesium oxide, basic slag (calcium silicate), and oyster shells. These products increase the pH of soils through various acid-base reactions. Calcium silicate neutralizes active acidity in the soil by reacting with H^+ ions to form monosilicic acid (H_4SiO_4), a neutral solute.

Decreasing the pH of Alkaline Soil

The pH of an alkaline soil can be reduced by adding acidifying agents or acidic organic materials. Elemental sulfur (90–99% S) has been used at application rates of 300–500 kg/ha – it slowly oxidizes in soil to form sulfuric acid. Acidifying fertilizers, such as ammonium sulfate, ammonium nitrate and urea, can help to reduce the pH of a soil because ammonium oxidises to form nitric acid. Acidifying organic materials include peat or sphagnum peat moss.

However, in high-pH soils with a high calcium carbonate content (more than 2%), it can be very costly and/or ineffective to attempt to reduce the pH with acids. In such cases, it is often more efficient to add phosphorus, iron, manganese, copper and/or zinc instead, because deficiencies of these nutrients are the most common reasons for poor plant growth in calcareous soils.

SOIL ACIDIFICATION

Soil acidification is the buildup of hydrogen cations, which reduces the soil pH. Chemically, this happens when a proton donor gets added to the soil. The donor can be an acid, such as nitric acid, sulfuric acid, or carbonic acid. It can also be a compound such as aluminium sulfate, which reacts in the soil to release protons. Acidification also occurs when base cations such as calcium, magnesium, potassium and sodium are leached from the soil.

Soil acidification naturally occurs as lichens and algae begin to break down rock surfaces. Acids continue with this dissolution as soil develops. With time and weathering, soils become more acidic in natural ecosystems. Soil acidification rates can vary, and increase with certain factors such as acid rain, agriculture and pollution.

Causes

Acid Rain

Rainfall is naturally acidic due to carbonic acid forming from carbon dioxide in the atmosphere. This compound causes rainfall pH to be around 5.0-5.5. When rainfall has a lower pH than natural levels, it can cause rapid acidification of soil. Sulfur dioxide and nitrogen oxides are precursors of stronger acids that can lead to acid rain production when they react with water in the atmosphere. These gases may be present in the atmosphere due to natural sources such as lightning and volcanic eruptions, or from anthropogenic emissions. Basic cations like calcium are leached from the soil as acidic rainfall flows, which allows aluminum and proton levels to increase.

Biological Weathering

Plant roots acidify soil by releasing protons and organic acids so as to chemically weather soil minerals. Decaying remains of dead plants on soil may also form organic acids which contribute to soil acidification. Acidification from leaf litter on soil is more pronounced under coniferous trees such as pine, spruce and fir, which return fewer base cations to the soil, than under deciduous trees.

Parent Materials

Certain parent materials also contribute to soil acidification. Granites and their allied igneous rocks are called "acidic" because they have a lot of free quartz, which produces silicic acid on weathering. Also, they have relatively low amounts of calcium and magnesium. Some sedimentary rocks such as shale and coal are rich in sulfides, which, when hydrated and oxidized, produce sulfuric acid which is much stronger than silicic acid. Many coal soils are too acidic to support vigorous plant growth, and coal gives off strong precursors to acid rain when it is burned. Marine clays are also sulfide-rich in many cases, and such clays become very acidic if they are drained to an oxidizing state.

Soil Amendments

Soil amendments such as fertilizers and manures can cause soil acidification. Sulfur based fertilizers can be highly acidifying, examples include elemental sulfur and iron sulfate while others like potassium sulfate have no significant effect on soil pH. While most nitrogen fertilizers have an acidifying effect, ammonium-based nitrogen fertilizers are more acidifying than other nitrogen sources. Ammonia-based nitrogen fertilizers include ammonium sulfate, diammonium phosphate, monoammonium phosphate, and ammonium nitrate. Organic nitrogen sources, such as urea and compost, are less acidifying. Nitrate sources which have little or no ammonium, such as calcium nitrate, magnesium nitrate, potassium nitrate, and sodium nitrate, are not acidifying.

Pollution

Acidification may also occur from nitrogen emissions into the air, as the nitrogen may end up deposited into the soil. Animal livestock is responsible for 64 percent of man-made ammonia emissions.

Anthropogenic sources of sulfur dioxides and nitrogen oxides are a large contributor to an increase in acid rain production. The use of fossil fuels and motor exhaust are the largest anthropogenic contributors to sulfuric gases and nitrogen oxides, respectively.

Effects

Soil acidification can cause damage to plants and organisms in the soil. In plants, soil acidification results in smaller, less durable roots. Acidic soils sometimes damage the root tips restricting further growth. Plant height is impaired and seed germination also decreases. Soil acidification impacts plant health, resulting in reduced cover and lower plant density. Soil acidification is directly linked to a decline in endangered species of plants.

In the soil, acidification reduces microbial and macrofaunal diversity. This can reduce soil structure decline which makes it more sensitive to erosion. There are less nutrients available in the soil, larger impact of toxic elements to plants, and consequences to soil biological functions (such as nitrogen fixation).

At a larger scale, soil acidification is linked to losses in agricultural productivity due to these effects.

Prevention and Management

Soil acidification is a common issue in long-term crop production which can be reduced by lime application. In soybean and corn crops grown in acidic soils, lime application resulted in nutrient restoration, increase in soil pH, increase in root biomass, and better plant health.

Different management strategies may also be applied to prevent further acidification: using less acidifying fertilizers, considering fertilizer amount and application timing to reduce nitrate-nitrogen leaching, good irrigation management with acid-neutralizing water, and considering the ratio of basic nutrients to nitrogen in harvested crops. Sulfur fertilizers should only be used in responsive crops, with a high rate of crop recovery.

Through reduction of anthropogenic sources of sulfur dioxides, nitrogen oxides, and with air-pollution control measures, acid rain and soil acidification have both decreased in northeastern American and eastern Canadian forests.

SOIL SALINITY

Soil salinity is the salt content in the soil; the process of increasing the salt content is known as salinization. Salts occur naturally within soils and water. Salination can be caused by natural processes such as mineral weathering or by the gradual withdrawal of an ocean. It can also come about through artificial processes such as irrigation and road salt.

Visibly salt-affected soils on rangeland in Colorado. Salts dissolved from the soil accumulate at the soil surface and are deposited on the ground and at the base of the fence post.

Saline incrustation in a PVC irrigation pipe from Brazil.

Natural Occurrence

Salts are a natural component in soils and water. The ions responsible for salination are: Na^+, K^+, Ca^{2+}, Mg^{2+} and Cl^-. As the Na^+ (sodium) predominates, soils can become sodic. Sodic soils present particular challenges because they tend to have very poor structure which limits or prevents water infiltration and drainage.

Over long periods of time, as soil minerals weather and release salts, these salts are flushed or leached out of the soil by drainage water in areas with sufficient precipitation. In addition to mineral weathering, salts are also deposited via dust and precipitation. In dry regions salts may accumulate, leading to naturally saline soils. This is the case, for example, in large parts of Australia. Human practices can increase the salinity of soils by the addition of salts in irrigation water. Proper irrigation management can prevent salt accumulation by providing adequate drainage water to leach added salts from the soil. Disrupting drainage patterns that provide leaching can also result in salt accumulations. An example of this occurred in Egypt in 1970 when the Aswan High Dam was built. The change in the level of ground water before the construction had enabled soil erosion, which led to high concentration of salts in the water table. After the construction, the continuous high level of the water table led to the salination of the arable land.

Dry Land Salinity

Salinity in drylands can occur when the water table is between two and three metres from the surface of the soil. The salts from the groundwater are raised by capillary action to the surface of the soil. This occurs when groundwater is saline (which is true in many areas), and is favored by land use practices allowing more rainwater to enter the aquifer than it could accommodate. For example, the clearing of trees for agriculture is a major reason for dryland salinity in some areas, since deep rooting of trees has been replaced by shallow rooting of annual crops.

Salinity Due to Irrigation

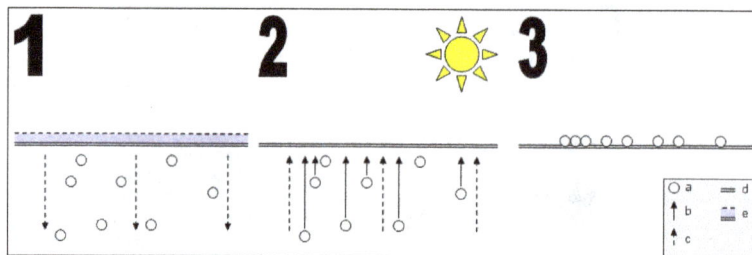

Rain or irrigation, in the absence of leaching, can bring salts to the surface by capillary action.

Salinity from irrigation can occur over time wherever irrigation occurs, since almost all water (even natural rainfall) contains some dissolved salts. When the plants use the water, the salts are left behind in the soil and eventually begin to accumulate. Since soil salinity makes it more difficult for plants to absorb soil moisture, these salts must be leached out of the plant root zone by applying additional water. This water in excess of plant needs is called the leaching fraction. Salination from irrigation water is also greatly increased by poor drainage and use of saline water for irrigating agricultural crops.

Salinity in urban areas often results from the combination of irrigation and groundwater processes. Irrigation is also now common in cities (gardens and recreation areas).

Consequences of Salinity

The consequences of salinity are:

- Detrimental effects on plant growth and yield.

- Damage to infrastructure (roads, bricks, corrosion of pipes and cables).

- Reduction of water quality for users, sedimentation problems, increased leaching of metals, especially copper, cadmium, manganese and zinc.

- Soil erosion ultimately, when crops are too strongly affected by the amounts of salts.

- More energy required to desalinate.

Salinity is an important land degradation problem. Soil salinity can be reduced by leaching soluble salts out of soil with excess irrigation water. Soil salinity control involves watertable control and flushing in combination with tile drainage or another form of subsurface drainage. A comprehensive treatment of soil salinity is available from the United Nations Food and Agriculture Organization.

Salt Tolerance of Crops

High levels of soil salinity can be tolerated if salt-tolerant plants are grown. Sensitive crops lose their vigor already in slightly saline soils, most crops are negatively affected by (moderately) saline soils, and only salinity-resistant crops thrive in severely saline soils. The University of Wyoming and the Government of Alberta report data on the salt tolerance of plants.

Field data in irrigated lands, under farmers' conditions, are scarce, especially in developing countries. However, some on-farm surveys have been made in Egypt, India, and Pakistan. Some examples are shown in the following gallery, with crops arranged from sensitive to very tolerant.

Graphs of crop yield and soil salinity in farmers' fields ordered by increasing salt tolerance.

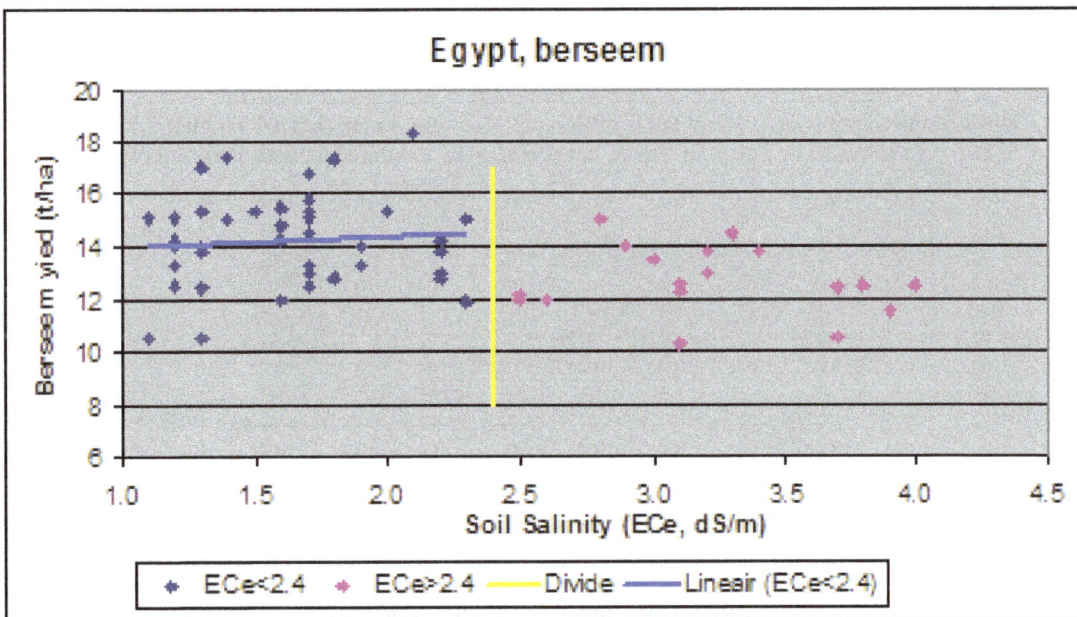

Berseem (clover), cultivated in Egypt's Nile Delta, is a salt-sensitive crop and tolerates an ECe value up to 2.4 dS/m, whereafter yields start to decline.

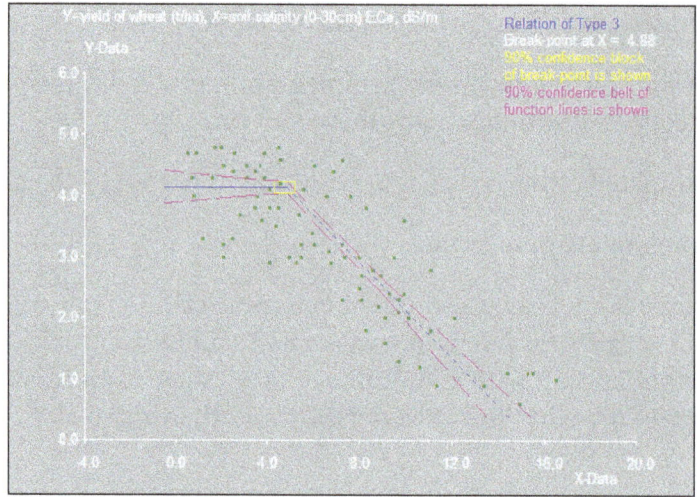

Wheat grown in Sampla, Haryana, India, is slightly
sensitive, tolerating an ECe value of 4.9 dS/m.

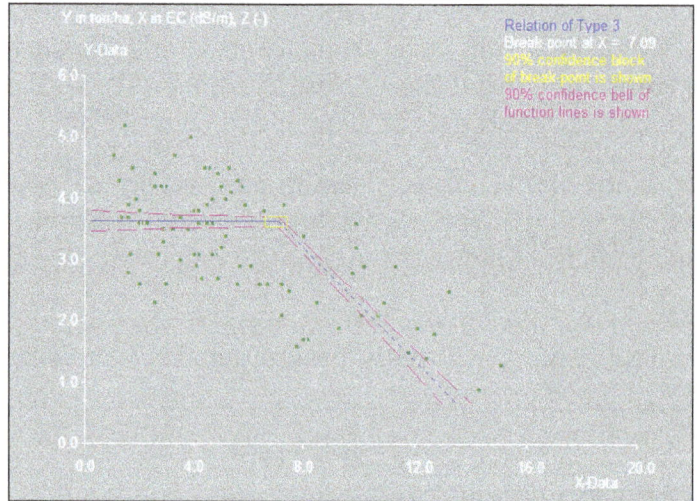

The field measurements in wheat fields in Gohana, Haryana, India, showed a higher tolerance
level of ECe = 7.1 dS/m. (The Egyptian wheat, not shown here, exhibited a tolerance point of 7.8 dS/m).

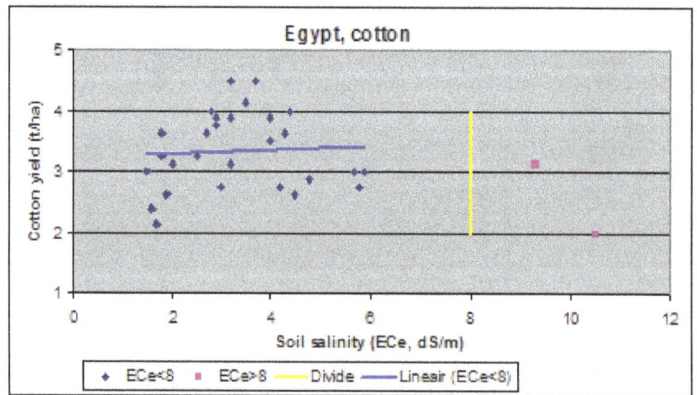

The cotton grown in the Nile Delta can be called salt-tolerant, with a critical ECe value
of 8.0 dS/m. However, due to scarcity of data beyond 8 dS/m, the maximum tolerance level
cannot be precisely determined and may actually be higher than that.

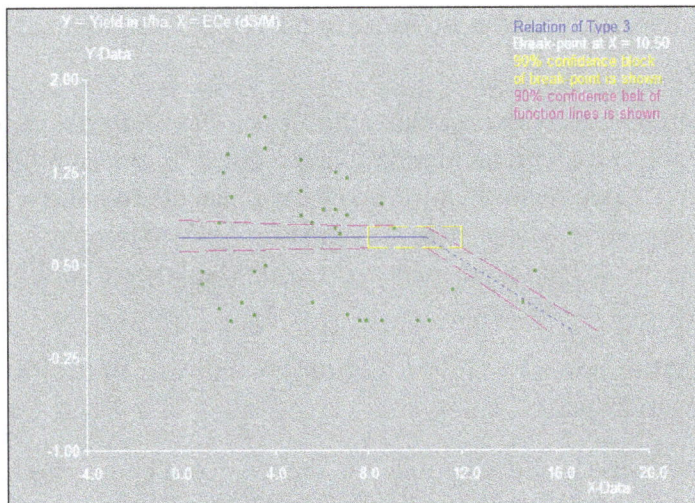

Sorghum from Khairpur, Pakistan, is quite tolerant;
it grows well up to ECe = 10.5 dS/m.

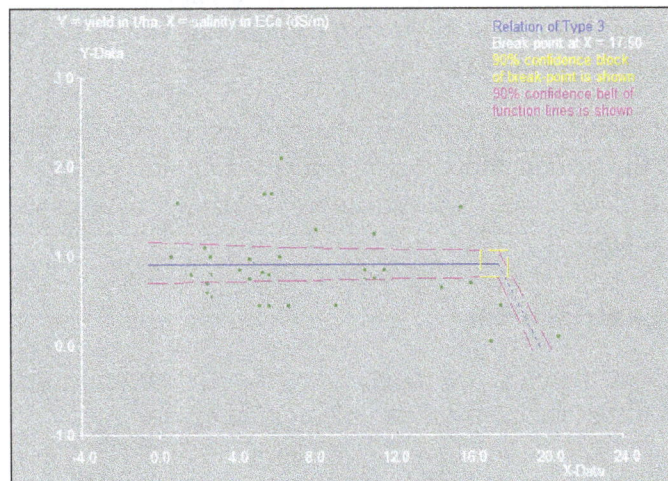

Cotton from Khairpur, Pakistan, is very tolerant; it grows well up to ECe = 15.5 dS/m.

SOIL ORGANIC MATTER

Organic matter is widely regarded as a vital component of a healthy soil. It is an important part of soil physical, chemical and biological fertility. Soil organic matter comprises all living soil organisms and all the remains of previous living organisms in their various degrees of decomposition. The living organisms can be animals, plants or micro-organisms, and can range in size from small animals to single cell bacteria only a few microns long.

Non-living organic matter can be considered to exist in four distinct pools:

- Organic matter dissolved in soil water.

- Particulate organic matter ranging from recently added plant and animal debris to partially decomposed material less than 50 microns in size, but all with an identifiable cell structure.

Particulate organic matter can constitute from a few percent up to 25% of the total organic matter in a soil.

- Humus which comprises both organic molecules of identifiable structure like proteins and cellulose, and molecules with no identifiable structure (humic and fulvic acids and humin) but which have reactive regions which allow the molecule to bond with other mineral and organic soil components. These molecules are moderate to large in size (molecular weights of 20,000 - 100,000). Humus usually represents the largest pool of soil organic matter, comprising over 50% of the total.

- Inert organic matter or charcoal derived from the burning of plants. Can be up to 10% of the total soil organic matter.

When plant and animal debris is added to soil, it is broken down by macro- and micro-organisms, initially into particulate organic matter, and finally into humus. The raw materials can vary greatly in their resistance to breakdown. Woody organic substances like lignins are very resistant, while more simple compounds like sugars are readily utilised. Along the way, microbial populations increase. In the process they synthesise their own compounds which add to the diversity. In turn, these organisms die and are consumed by others.

Carbon dioxide is a by-product of this complex chain of processes (microbes breathe out CO_2 just like we do). Over half of the carbon added to soil is lost as CO_2 during breakdown. Because of their varying reactivity, the turnover times for these different carbon fractions varies from a few months to tens of thousands of years.

Measuring Soil Organic Matter

While living organisms, particularly the plants we grow, are of vital importance to us, it is the non-living organic matter that we measure as 'soil organic matter'.

The most common methods for measuring soil organic matter in current use actually measure the amount of carbon in the soil. This is done by oxidising the carbon and measuring either the amount of oxidant used (wet oxidation, usually using dichromate) or the CO_2 given off in the process (combustion method with specific detection).

Laboratories these days generally report results as soil organic carbon. Those that report as soil organic matter have usually measured carbon and converted to organic matter by multiplying by 1.72. However, this conversion factor is not the same for all soils, and it is more precise to report soil carbon rather than organic matter.

The Amount of Organic Matter in Soil

The amount of carbon (the measure of organic matter) in a soil depends on a range of factors, and reflects the balance between accumulation and breakdown. The main factors are:

- Climate - For similar soils under similar management, carbon is greater in areas of higher rainfall, and lower in areas of higher temperature. The rate of decomposition doubles for every 8 or 9 °C increase in mean annual temperature.

- Soil type - Clay helps protect organic matter from breakdown, either by binding organic matter strongly or by forming a physical barrier which limits microbial access. Clay soils in the same area under similar management will tend to retain more carbon than sandy soils. Hence the sandy sodosols of the northern midlands have less carbon than the clay loam ferrosols of the north west regardless of management.

- Vegetative growth - The more vegetative production the greater are the inputs of carbon. Also, the more woody this vegetation is (greater C:N ratio), the slower it will breakdown. So, the crop system can strongly affect carbon concentrations.

- Topography - Soils at the bottom of slopes generally have higher carbon because these areas are generally wetter and have higher clay contents. Poorly drained areas have much slower rates of carbon breakdown.

- Tillage - Tillage will increase carbon breakdown. However, the impact of tillage is generally outweighed by the effect of management on the amount of carbon grown and returned to the soil. An exception to this is where tillage leads to increased erosion.

Soil (depth)	Pasture	Intermittent cropping	Frequent cropping
Ferrosol (0-150 mm) Red soil on basalt	6.4	4.9	3.8
Dermosol (0-75 mm) Cressy clay loam	7.0	4.3	4.2
Sodosol (0-150 mm) Sandy soil over clay	2.7	2.3	1.8

Benefits of Organic Matter

Organic matter can be considered a pivotal component of the soil because of its role in physical, chemical and biological processes. Many of these functions interact. For example, the high cation exchange properties of organic matter are a major means by which organic matter is able to bind soil particles together in a more stable structure. The reactive regions present in humus are numerous, and give these molecules a capacity to bind to each other and to mineral soil particles, and also to react with cations (positive charge, e.g. $Ca2+$, $K+$) in the soil solution.

The density of cation exchange capacity (CEC) of organic matter is greater than it is for clay minerals. While a high CEC is an important attribute of soil organic matter, please note that organic matter does not have an anion (negative) exchange capacity, and is therefore not able to bind anions like phosphate and sulphate. However, organic matter is a substantial reservoir for phosphorus and sulphur, as well as nitrogen. These elements are bound within the organic structure, and are released to the soil solution when microbes break down organic matter.

The ratio of carbon:nitrogen:sulphur:phosphorus in organic matter is roughly 100:10:1.5:1.5. A hectare of soil 10 cm deep with a bulk density of 1 tonne/m3 weighs 1,000,000 kg. Therefore, soil with a carbon content of 3% would contain 3,000 kg of organic nitrogen, and 450 kg each of organic phosphorus and sulphur per hectare. Not all of this is mineralised each year, but there is considerable potential for nutrients in organic matter to contribute to plant requirements. These should be taken into account, particularly the nitrogen.

Table: Functions of soil organic matter.

Physical functions	Chemical functions	Biological functions
• Bind soil particles together in stable aggregates • Influence water holding and aeration • Influence soil temperature	• Major source of cation exchange capacity • Source of pH buffering • Binding site for heavy metals and pesticides	• Food source for microbes and small animals • Major reservoir of plant nutrients

Table: Cation exchange capacity of different soil particles.

Soil particle	CEC (cmol/kg)
Humus	100-300
Smectites (black swelling clays)	60-150
Kaolinite (white potter's clay)	2-15
Iron and aluminium oxides (from ferrosols)	<1

References

- Soil-fertility-its-meaning-causes-and-maintenance-with-diagram, soil: biologydiscussion.com, Retrieved 14 August, 2019

- Pezeshki, S. R.; delaune, R. D. (2012-07-26). "Soil Oxidation-Reduction in Wetlands and Its Impact on Plant Functioning". Biology. 1 (2): 196–221. Doi:10.3390/biology1020196. PMC 4009779. PMID 24832223

- Conen F, and Smith K. (2000) An explanation of linear increases in gas concentration under closed chambers used to measure gas exchange between soil and the atmosphere. European Journal of Soil Science. 51, 111-117

- Berg, B., mcclaugherty, C., 2007. Plant litter: decomposition, humus formation, carbon sequestration, 2nd ed. Springer, 338 pp., ISBN 3-540-74922-5

- R.F. Isbell and the National Committee on Soil and Terrain (2016). "Australian Soil Classification, second edition (as Online Interactive Key)". CSIRO. Retrieved 11 February 2016

- Timbal, B., Power, S., Colman, R., Viviand, J., & Lirola, S., 2002. "Does Soil Moisture Influence Climate Variability and Predictability over Australia?" Archived 2014-02-13 at the Wayback Machine Journal of Climate, Volume 15, pp.1230 – 1238. Viewed May 2007

Soil Management

The application of practices, operations, and treatments for the protection of soil and enhancement of its performance, such as soil fertility or soil mechanics, is known as soil management. Some of the commonly used practices and techniques of soil management are crop rotation, nutrient management, tillage, green manuring, cover crop etc. The different practices and techniques of soil management have been thoroughly discussed in this chapter.

Soil management is an integral part of land management and may focus on differences in soil types and soil characteristics to define specific interventions that are aimed to enhance the soil quality for the land use selected. Specific soil management practices are needed to protect and conserve the soil resources. Specific interventions also exist to enhance the carbon content in soils in order to mitigate climate change.

Furthermore, some practices and treatments that involve Other Sustainable Soil Management Tools can benefit the soil.

Close-up of rice terraces, Madagascar.

The goal of good soil management is to meet essential plant needs. Healthy plants need water, nutrients, oxygen, and a physical medium that allows seeds to germinate, shoots to emerge and grow up toward the sunlight, and roots to anchor the plant by growing strong and deep.

In the process of growing crops on farms, we use a number of farming practices to manage soil in the field. These include tilling, cultivating, adding fertilizers and lime, growing cover crops, applying compost or manure, rotating crops, and other practices. Many years of agricultural research have shown us that how and when we use these practices makes a big difference to the quality of our soils. When we use these practices correctly, we can improve soil fertility, soil physical structure, and biological activity, and also protect soils from erosion. Soils that are properly managed for soil quality produce healthier, higher-yielding crops.

Moldboard plowing, shown here, is a type of primary tillage.

Managing soil organic matter is extremely important because organic matter plays a role in almost all aspects of soil quality. Soil organic matter is composed of plant and animal residues, and substances produced by decomposition. Most agricultural soils contain only a small proportion of organic matter, usually less than 5%, but even that small amount strongly influences the vital functions of soil. Soil organic matter releases plant nutrients as it decomposes, thus improving soil fertility. Soil organic matter improves the structure of soils by improving water holding capacity, promoting the aggregation of soil particles, and helping to keep the soil aerated so roots can grow easily. Organic matter feeds the microorganisms and other living organisms in the soil, promoting biological activity and helping to fight pests. Practices such as using cover crops, applying manure and compost, rotating crops, and controlling erosion for soil conservation, can maintain or increase soil organic matter. Other practices, especially plowing, tilling and cultivating, can decrease the amount of organic matter in the soil.

Soil management practices are used to improve crop production, but each can affect soil quality, as well.

SOIL SURVEY

Soil survey is the study and mapping of soils in the fields. It is starting point for all soil researches.

The following are the main objectives of soil survey:

- To describe and classify soils giving uniform system of classification with uniform nomenclature in order to correlate the soils of different area.

- To show distribution of different soils in the field (soil mapping).

- To provide data for making interpretations as to the adaptability of particular soils for agricultural purpose and also for many other purposes (as in soil management).

Soil surveys are of the following three types:

- Detailed Soil Survey: In this, soil boundaries are plotted accurately on maps on the basis of observations made throughout the surveyed area. In this, geographical distribution of soil is also described. Detailed soil surveys are important in the sense that they provide information's needed for planning land use and management and formulating agricultural research and extension programmes.

- Reconnaissance Survey: In this, soil boundaries are plotted from the observations made at intervals.

- Detailed Reconnaissance Survey: In this, a part of surveyed area is plotted on the map by detailed method and remainder by reconnaissance method. In mapping of soils either aerial photographs are taken or good topographic maps are made. In mapping of cultivated areas, generally a scale of four inches to a mile is used but detailed or special maps are made on scales of six inches, eight inches or twelve inches to a mile.

In India, soil surveys are being carried out by central and provincial agencies. Some surveys of limited areas have been conducted for specific objectives, such as fertility surveys, surveys for soil classification, survey from geological point of view, for physicochemical properties of surface sample and surveys for genetic classification.

Pre and post-irrigation surveys have also been used for limited areas under command of big irrigation projects to determine the nature, quality and the concentration of soluble salts in different horizons of soil.

Aerial Photo-interpretation for Soil Surveys

Normally the scales recommended for detailed survey are 1: 10,000 or 1: 25,000. In such surveys, the air photos serve as an ideal base for plotting soil boundaries. This is on account of the wealth of pictorial detail available in the air photo, thus rendering orientation and navigation very easy. Normally a limited use of pocket stereoscope is made for delineating boundaries based on relief and other features. The final soil map in such surveys as well as the base photo are prepared on large scale.

Medium and Small Scale Map

Aerial photo-interpretation is of greater utility in this case. The degree of utility increases from medium to small scale. The procedure for photo-interpretation is based mainly on the principle that physiographic units have unique soil patterns. The scale recommended for semi- detailed reconnaissance surveys are 1: 25,000 to 1: 1, 00,000 and 1: 2, 50,000. In similarity to the techniques used by geologists and foresters, the soil scientists also make use of such elements as relief, slope, geological features, vegetation, tone, textures, pattern etc.

Systematic Photo-interpretation

The most favored method adopted by soil scientists involves delineation of the soil boundaries by air photos which are sub-divisions of physiographic units. In the next stage, they select sample areas that represent all the delineations of the photo-interpretation.

These consist of long, narrow areas which cut across as many sub-divisions as possible. Detailed soil observations are taken up in each of the delineation of the interpretation. Based upon the number of such observations, the soils of each unit are classified and related to the photo-interpretation unit. Thus soil composition of each unit is established.

With the finding of the sample areas as basic, the remaining aerial photographs are interpreted by the principle of extrapolation of knowledge gained in the sample area studies. It may be mentioned that this "adjusted photo-interpretation" is very often a partial revision of the pre-field interpretation done in the office. Final phase comprises a "selective ground check". This achieves the testing of the validity of the adjusted photo-interpretation. The resulting field sheet namely air photos contain the correct soil boundaries.

SOIL MECHANICS

Soil mechanics is a branch of soil physics and applied mechanics that describes the behavior of soils. It differs from fluid mechanics and solid mechanics in the sense that soils consist of a heterogeneous mixture of fluids (usually air and water) and particles (usually clay, silt, sand, and gravel) but soil may also contain organic solids and other matter. Along with rock mechanics, soil mechanics provides the theoretical basis for analysis in geotechnical engineering, a subdiscipline of civil engineering, and engineering geology, a subdiscipline of geology. Soil mechanics is used to analyze the deformations of and flow of fluids within natural and man-made structures that are supported on or made of soil, or structures that are buried in soils. Example applications are building and bridge foundations, retaining walls, dams, and buried pipeline systems. Principles of soil mechanics are also used in related disciplines such as engineering geology, geophysical engineering, coastal engineering, agricultural engineering, hydrology and soil physics.

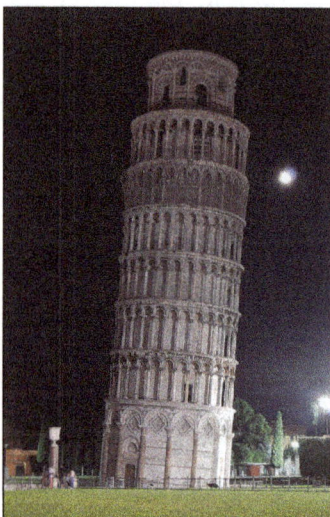

The Leaning Tower of Pisa– an example of a problem due to deformation of soil.

Fox Glacier, New Zealand: Soil produced and transported by intense weathering and erosion.

Slope instability issues for a temporary flood control levee in North Dakota.

Earthwork in Germany.

Genesis and Composition of Soils

Genesis

The primary mechanism of soil creation is the weathering of rock. All rock types (igneous rock, metamorphic rock and sedimentary rock) may be broken down into small particles to create soil. Weathering mechanisms are physical weathering, chemical weathering, and biological weathering Human activities such as excavation, blasting, and waste disposal, may also create soil. Over geologic time, deeply buried soils may be altered by pressure and temperature to become metamorphic or sedimentary rock, and if melted and solidified again, they would complete the geologic cycle by becoming igneous rock.

Physical weathering includes temperature effects, freeze and thaw of water in cracks, rain, wind, impact and other mechanisms. Chemical weathering includes dissolution of matter composing a rock and precipitation in the form of another mineral. Clay minerals, for example can be formed by weathering of feldspar, which is the most common mineral present in igneous rock.

The most common mineral constituent of silt and sand is quartz, also called silica, which has the chemical name silicon dioxide. The reason that feldspar is most common in rocks but silica is more prevalent in soils is that feldspar is much more soluble than silica.

Silt, Sand, and Gravel are basically little pieces of broken rocks.

According to the Unified Soil Classification System, silt particle sizes are in the range of 0.002 mm to 0.075 mm and sand particles have sizes in the range of 0.075 mm to 4.75 mm.

Gravel particles are broken pieces of rock in the size range 4.75 mm to 100 mm. Particles larger than gravel are called cobbles and boulders.

Transport

Soil deposits are affected by the mechanism of transport and deposition to their location. Soils that are not transported are called residual soils—they exist at the same location as the rock from which they were generated. Decomposed granite is a common example of a residual soil. The common mechanisms of transport are the actions of gravity, ice, water, and wind. Wind blown soils include dune sands and loess. Water carries particles of different size depending on the speed of the water, thus soils transported by water are graded according to their size. Silt and clay may settle out in a lake, and gravel and sand collect at the bottom of a river bed. Wind blown soil deposits (aeolian soils) also tend to be sorted according to their grain size. Erosion at the base of glaciers is powerful enough to pick up large rocks and boulders as well as soil; soils dropped by melting ice can be a well graded mixture of widely varying particle sizes. Gravity on its own may also carry particles down from the top of a mountain to make a pile of soil and boulders at the base; soil deposits transported by gravity are called colluvium.

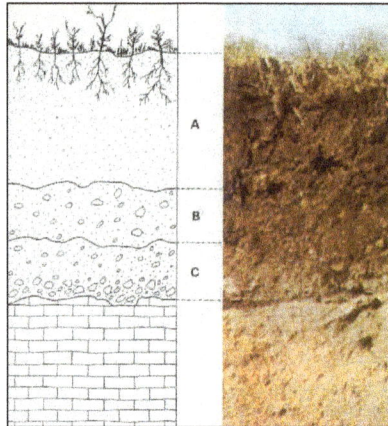

Soil horizons: a) top soil and colluvium b) mature residual soil c) young residual soil d) weathered rock.

The mechanism of transport also has a major effect on the particle shape. For example, low velocity grinding in a river bed will produce rounded particles. Freshly fractured colluvium particles often have a very angular shape.

Soil Composition

Soil Mineralogy

Silts, sands and gravels are classified by their size, and hence they may consist of a variety of minerals. Owing to the stability of quartz compared to other rock minerals, quartz is the most common constituent of sand and silt. Mica, and feldspar are other common minerals present in sands and silts. The mineral constituents of gravel may be more similar to that of the parent rock.

The common clay minerals are montmorillonite or smectite, illite, and kaolinite or kaolin. These minerals tend to form in sheet or plate like structures, with length typically ranging between 10^{-7} m and 4×10^{-6} m and thickness typically ranging between 10^{-9} m and 2×10^{-6} m, and they have a relatively large specific surface area. The specific surface area (SSA) is defined as the ratio of the surface area of particles to the mass of the particles. Clay minerals typically have specific surface areas in the range of 10 to 1,000 square meters per gram of solid. Due to the large surface area available for chemical, electrostatic, and van der Waals interaction, the mechanical behavior of clay minerals is very sensitive to the amount of pore fluid available and the type and amount of dissolved ions in the pore fluid. To anticipate the effect of clay on the way a soil will behave, it is necessary to know the kinds of clays as well as the amount present. As home builders and highway engineers know all too well, soils containing certain high-activity clays make very unstable material on which to build because they swell when wet and shrink when dry. This shrink-and-swell action can easily crack foundations and cause retaining walls to collapse. These clays also become extremely sticky and difficult to work with when they are wet. In contrast, low-activity clays, formed under different conditions, can be very stable and easy to work with.

The minerals of soils are predominantly formed by atoms of oxygen, silicon, hydrogen, and aluminum, organized in various crystalline forms. These elements along with calcium, sodium, potassium, magnesium, and carbon constitute over 99 per cent of the solid mass of soils.

Grain Size Distribution

Soils consist of a mixture of particles of different size, shape and mineralogy. Because the size of the particles obviously has a significant effect on the soil behavior, the grain size and grain size distribution are used to classify soils. The grain size distribution describes the relative proportions of particles of various sizes. The grain size is often visualized in a cumulative distribution graph which, for example, plots the percentage of particles finer than a given size as a function of size. The median grain size, D_{50}, is the size for which 50% of the particle mass consists of finer particles. Soil behavior, especially the hydraulic conductivity, tends to be dominated by the smaller particles, hence, the term "effective size", denoted by D_{10}, is defined as the size for which 10% of the particle mass consists of finer particles.

Sands and gravels that possess a wide range of particle sizes with a smooth distribution of particle sizes are called *well graded* soils. If the soil particles in a sample are predominantly in a relatively narrow range of sizes, the sample is *uniformly graded*. If a soil sample has distinct gaps in the gradation curve, e.g., a mixture of gravel and fine sand, with no coarse sand, the sample may be *gap graded*. *Uniformly graded* and *gap graded* soils are both considered to be *poorly graded*. There are many methods for measuring particle-size distribution. The two traditional methods are sieve analysis and hydrometer analysis.

Sieve Analysis

Sieve.

The size distribution of gravel and sand particles are typically measured using sieve analysis. A stack of sieves with accurately dimensioned holes between a mesh of wires is used to separate the particles into size bins. A known volume of dried soil, with clods broken down to individual particles, is put into the top of a stack of sieves arranged from coarse to fine. The stack of sieves is shaken for a standard period of time so that the particles are sorted into size bins. This method works reasonably well for particles in the sand and gravel size range. Fine particles tend to stick to each other, and hence the sieving process is not an effective method. If there are a lot of fines (silt and clay) present in the soil it may be necessary to run water through the sieves to wash the coarse particles and clods through.

A variety of sieve sizes are available. The boundary between sand and silt is arbitrary. According to the Unified Soil Classification System, a number 4 sieve (4 openings per inch) having 4.75 mm opening size separates sand from gravel and a number 200 sieve with an 0.075 mm opening separates sand from silt and clay. According to the British standard, 0.063 mm is the boundary between sand and silt, and 2 mm is the boundary between sand and gravel.

Hydrometer Analysis

The classification of fine-grained soils, i.e., soils that are finer than sand, is determined primarily by their Atterberg limits, not by their grain size. If it is important to determine the grain size distribution of fine-grained soils, the hydrometer test may be performed. In the hydrometer tests, the soil particles are mixed with water and shaken to produce a dilute suspension in a glass cylinder, and then the cylinder is left to sit. A hydrometer is used to measure the density of the suspension as a function of time. Clay particles may take several hours to settle past the depth of measurement of the hydrometer. Sand particles may take less than a second. Stoke's law provides the theoretical basis to calculate the relationship between sedimentation velocity and particle size. ASTM provides the detailed procedures for performing the Hydrometer test.

Clay particles can be sufficiently small that they never settle because they are kept in suspension by Brownian motion, in which case they may be classified as colloids.

Mass-volume Relations

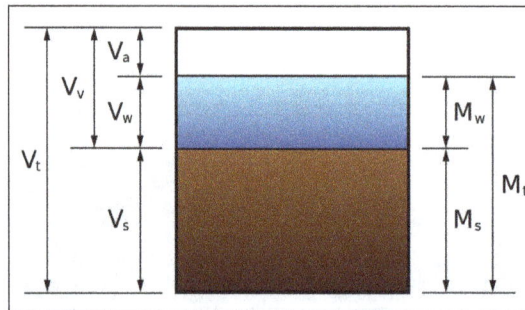

A phase diagram of soil indicating the masses and volumes of air, solid, water, and voids.

There are a variety of parameters used to describe the relative proportions of air, water and solid in a soil. This section defines these parameters and some of their interrelationships. The basic notation is as follows:

- V_a, V_w, and V_s represent the volumes of air, water and solids in a soil mixture;

- W_a, W_w, and W_s represent the weights of air, water and solids in a soil mixture;

- M_a, M_w, and M_s represent the masses of air, water and solids in a soil mixture;

- ρ_a, ρ_w, and ρ_s represent the densities of the constituents (air, water and solids) in a soil mixture.

The weights, W, can be obtained by multiplying the mass, M, by the acceleration due to gravity, g; e.g., $W_s = M_s g$.

Specific Gravity is the ratio of the density of one material compared to the density of pure water ($\rho_w = 1g / cm^3$).

Specific gravity of solids, $G_s = \dfrac{\rho_s}{\rho_w}$

Specific weight, conventionally denoted by the symbol γ may be obtained by multiplying the density (ρ) of a material by the acceleration due to gravity, g.

Density, Bulk Density, or Wet Density, ρ, are different names for the density of the mixture, i.e., the total mass of air, water, solids divided by the total volume of air water and solids (the mass of air is assumed to be zero for practical purposes):

$$\rho = \frac{M_s + M_w}{V_s + V_w + V_a} = \frac{M_t}{V_t}$$

Dry Density, ρ_d, is the mass of solids divided by the total volume of air water and solids:

$$\rho_d = \frac{M_s}{V_s + V_w + V_a} = \frac{M_s}{V_t}$$

Buoyant Density, ρ', defined as the density of the mixture minus the density of water is useful if the soil is submerged under water:

$$\rho' = \rho - \rho_w$$

where ρ_w is the density of water.

Water Content, w is the ratio of mass of water to mass of solid. It is easily measured by weighing a sample of the soil, drying it out in an oven and re-weighing. Standard procedures are described by ASTM.

$$w = \frac{M_w}{M_s} = \frac{W_w}{W_s}$$

Void ratio, e, is the ratio of the volume of voids to the volume of solids:

$$e = \frac{V_v}{V_s} = \frac{V_v}{V_T - V_V} = \frac{n}{1-n}$$

Porosity, n, is the ratio of volume of voids to the total volume, and is related to the void ratio:

$$n = \frac{V_v}{V_t} = \frac{V_v}{V_s + V_v} = \frac{e}{1+e}$$

Degree of saturation, S, is the ratio of the volume of water to the volume of voids:

$$S = \frac{V_w}{V_v}$$

From the above definitions, some useful relationships can be derived by use of basic algebra.

$$\rho = \frac{(G_s + Se)\rho_w}{1+e}$$

$$\rho = \frac{(1+w)G_s\rho_w}{1+e}$$

$$w = \frac{Se}{G_s}$$

Soil Classification

Geotechnical engineers classify the soil particle types by performing tests on disturbed (dried, passed through sieves, and remolded) samples of the soil. This provides information about the characteristics of the soil grains themselves. Classification of the types of grains present in a soil does not account for important effects of the *structure* or *fabric* of the soil, terms that describe compactness of the particles and patterns in the arrangement of particles in a load carrying framework as well as the pore size and pore fluid distributions. Engineering geologists also classify soils based on their genesis and depositional history.

Classification of Soil Grains

In the US and other countries, the Unified Soil Classification System (USCS) is often used for soil classification. Other classification systems include the British Standard BS 5930 and the AASHTO soil classification system.

Classification of Sands and Gravels

In the USCS, gravels (given the symbol G) and sands (given the symbol S) are classified according to their grain size distribution. For the USCS, gravels may be given the classification symbol *GW* (well-graded gravel), *GP* (poorly graded gravel), *GM* (gravel with a large amount of silt), or *GC*(-gravel with a large amount of clay). Likewise sands may be classified as being *SW, SP, SM* or *SC*. Sands and gravels with a small but non-negligible amount of fines (5–12%) may be given a dual classification such as *SW-SC*.

Atterberg Limits

Clays and Silts, often called 'fine-grained soils', are classified according to their Atterberg limits; the most commonly used Atterberg limits are the Liquid Limit (denoted by *LL* or w_l), Plastic Limit (denoted by *PL* or w_p), and Shrinkage Limit (denoted by *SL*).

The Liquid Limit is the water content at which the soil behavior transitions from a plastic solid to a liquid. The Plastic Limit is the water content at which the soil behavior transitions from that of a plastic solid to a brittle solid. The Shrinkage Limit corresponds to a water content below which the soil will not shrink as it dries. The consistency of fine grained soil varies in proportional to the water content in a soil.

As the transitions from one state to another are gradual, the tests have adopted arbitrary definitions to determine the boundaries of the states. The liquid limit is determined by measuring the water content for which a groove closes after 25 blows in a standard test. Alternatively, a fall cone testapparatus may be use to measure the liquid limit. The undrained shear strength of remolded soil at the liquid limit is approximately 2 kPa. The Plastic Limit is the water content below which it is not possible to roll by hand the soil into 3 mm diameter cylinders. The soil cracks or breaks up as it is rolled down to this diameter. Remolded soil at the plastic limit is quite stiff, having an undrained shear strength of the order of about 200 kPa.

The Plasticity Index of a particular soil specimen is defined as the difference between the Liquid Limit and the Plastic Limit of the specimen; it is an indicator of how much water the soil particles in the specimen can absorb, and correlates with many engineering properties like permeability, compressibility, shear strength and others. Generally, the clay having high plasticity have lower permeability and also they are also difficult to be compacted.

Classification of Silts and Clays

According to the Unified Soil Classification System (USCS), silts and clays are classified by plotting the values of their plasticity index and liquid limiton a plasticity chart. The A-Line on the chart separates clays (given the USCS symbol *C*) from silts (given the symbol *M*). LL=50% separates high plasticity soils (given the modifier symbol *H*) from low plasticity soils (given the modifier symbol *L*). A soil that plots above the A-line and has LL>50% would, for example, be classified as *CH*. Other possible classifications of silts and clays are *ML*, *CL* and *MH*. If the Atterberg limits plot in the"hatched" region on the graph near the origin, the soils are given the dual classification 'CL-ML'.

Indices Related to Soil Strength

Liquidity Index

The effects of the water content on the strength of saturated remolded soils can be quantified by the use of the *liquidity index, LI*:

$$LI = \frac{w - PL}{LL - PL}$$

When the LI is 1, remolded soil is at the liquid limit and it has an undrained shear strength of about 2 kPa. When the soil is at the plastic limit, the LI is 0 and the undrained shear strength is about 200 kPa.

Relative Density

The density of sands (cohesionless soils) is often characterized by the relative density, D_r

$$D_r = \frac{e_{max} - e}{e_{max} - e_{min}} 100\%$$

where: e_{max} is the "maximum void ratio" corresponding to a very loose state, e_{min} is the "minimum void ratio" corresponding to a very dense state and e is the *in situ* void ratio. Methods used to calculate relative density are defined in ASTM D4254-00(2006).

Thus if $D_r = 100\%$ the sand or gravel is very dense, and if $D_r = 0\%$ the soil is extremely loose and unstable.

Seepage: Steady State Flow of Water

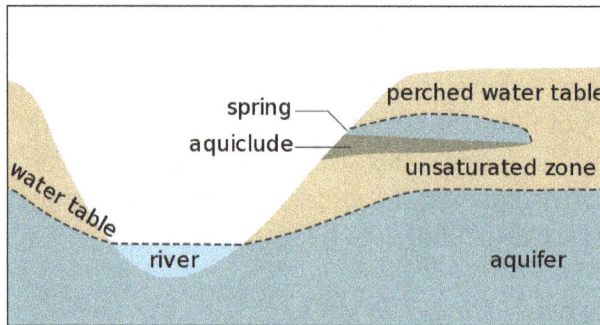

A cross section showing the water table varying with surface topography as well as a perched water table.

If fluid pressures in a soil deposit are uniformly increasing with depth according to $u = \rho_w g z_w$ then hydrostatic conditions will prevail and the fluids will not be flowing through the soil. z_w is the depth below the water table. However, if the water table is sloping or there is a perched water table as indicated in the accompanying sketch, then seepage will occur. For steady state seepage, the seepage velocities are not varying with time. If the water tables are changing levels with time, or if the soil is in the process of consolidation, then steady state conditions do not apply.

Darcy's Law

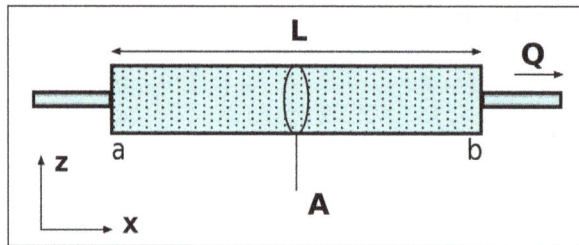

Diagram showing definitions and directions for Darcy's law.

Darcy's law states that the volume of flow of the pore fluid through a porous medium per unit time is proportional to the rate of change of excess fluid pressure with distance. The constant of proportionality includes the viscosity of the fluid and the intrinsic permeability of the soil. For the simple case of a horizontal tube filled with soil:

$$Q = \frac{-KA}{\mu} \frac{(u_b - u_a)}{L}$$

The total discharge, Q (having units of volume per time, e.g., ft³/s or m³/s), is proportional to the intrinsic permeability, K, the cross sectional area, A, and rate of pore pressure change with distance, $\dfrac{u_b - u_a}{L}$, and inversely proportional to the dynamic viscosity of the fluid, μ. The negative sign is needed because fluids flow from high pressure to low pressure. So if the change in pressure is negative (in the x-direction) then the flow will be positive (in the x-direction). The above equation works well for a horizontal tube, but if the tube was inclined so that point b was a different elevation than point a, the equation would not work. The effect of elevation is accounted for by replacing the pore pressure by *excess pore pressure*, u_e defined as:

$$u_e = u - \rho_w g z$$

where z is the depth measured from an arbitrary elevation reference (datum). Replacing u by u_e we obtain a more general equation for flow:

$$Q = \frac{-KA}{\mu} \frac{(u_{e,b} - u_{e,a})}{L}.$$

Dividing both sides of the equation by A, and expressing the rate of change of excess pore pressure as a derivative, we obtain a more general equation for the apparent velocity in the x-direction:

$$v_x = \frac{-K}{\mu} \frac{du_e}{dx}$$

Where $v_x = Q / A$ has units of velocity and is called the *Darcy velocity* (or the *specific discharge, filtration velocity*, or *superficial velocity*). The *pore* or *interstitial velocity* v_{px} is the average velocity of fluid molecules in the pores; it is related to the Darcy velocity and the porosity n through the *Dupuit-Forchheimer relationship*:

$$v_{px} = \frac{v_x}{n}$$

Civil engineers predominantly work on problems that involve water and predominantly work on problems on earth (in earth's gravity). For this class of problems, civil engineers will often write Darcy's law in a much simpler form:

$$v = ki$$

Where k is the hydraulic conductivity, defined as $k = \dfrac{K \rho_w g}{\mu_w}$, and i is the hydraulic gradient. The hydraulic gradient is the rate of change of total headwith distance. The total head, h at a point is defined as the height (measured relative to the datum) to which water would rise in a piezometer at that point. The total head is related to the excess water pressure by:

$$u_e = \rho_w g h + Constant$$

And the *Constant* is zero if the datum for head measurement is chosen at the same elevation as the origin for the depth, z used to calculate u_e.

Typical Values of Hydraulic Conductivity

Values of hydraulic conductivity, k, can vary by many orders of magnitude depending on the soil type. Clays may have hydraulic conductivity as small as about $10^{-12} \frac{m}{s}$, gravels may have hydraulic conductivity up to about $10^{-1} \frac{m}{s}$. Layering and heterogeneity and disturbance during the sampling and testing process make the accurate measurement of soil hydraulic conductivity a very difficult problem.

Flownets

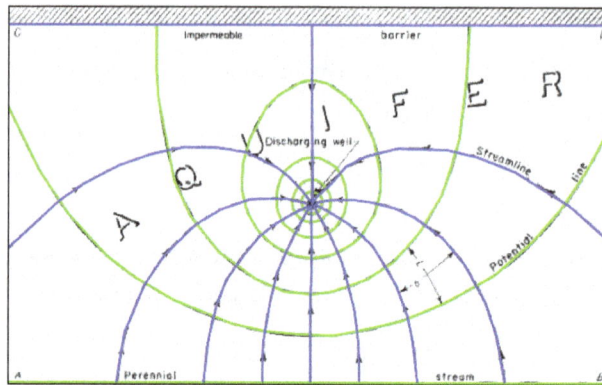

A plan flow net to estimate flow of water from a stream to a discharging well.

Darcy's Law applies in one, two or three dimensions. In two or three dimensions, steady state seepage is described by Laplace's equation. Computer programs are available to solve this equation. But traditionally two-dimensional seepage problems were solved using a graphical procedure known as flownet. One set of lines in the flownet are in the direction of the water flow (flow lines), and the other set of lines are in the direction of constant total head (equipotential lines). Flownets may be used to estimate the quantity of seepage under dams and sheet piling.

Seepage Forces and Erosion

When the seepage velocity is great enough, erosion can occur because of the frictional drag exerted on the soil particles. Vertically upwards seepage is a source of danger on the downstream side of sheet piling and beneath the toe of a dam or levee. Erosion of the soil, known as "soil piping", can lead to failure of the structure and to sinkhole formation. Seeping water removes soil, starting from the exit point of the seepage, and erosion advances upgradient. The term "sand boil" is used to describe the appearance of the discharging end of an active soil pipe.

Seepage Pressures

Seepage in an upward direction reduces the effective stress within the soil. When the water pressure at a point in the soil is equal to the total vertical stress at that point, the effective stress is zero and the soil has no frictional resistance to deformation. For a surface layer, the vertical effective stress becomes zero within the layer when the upward hydraulic gradient is equal to the critical gradient. At zero effective stress soil has very little strength and layers of relatively impermeable soil may heave up due to the underlying water pressures. The loss in strength due to upward seepage is a common contributor to levee failures. The condition of zero effective stress associated with upward seepage is also called liquefaction, quicksand, or a boiling condition. Quicksand was so named because the soil particles move around and appear to be 'alive' (the biblical meaning of 'quick' – as opposed to 'dead').

Effective Stress and Capillarity: Hydrostatic Conditions

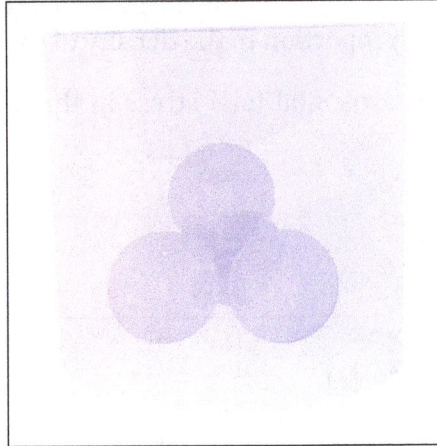

Spheres immersed in water, reducing effective stress.

To understand the mechanics of soils it is necessary to understand how normal stresses and shear stresses are shared by the different phases. Neither gas nor liquid provide significant resistance to shear stress. The shear resistance of soil is provided by friction and interlocking of the particles. The friction depends on the intergranular contact stresses between solid particles. The normal stresses, on the other hand, are shared by the fluid and the particles. Although the pore air is relatively compressible, and hence takes little normal stress in most geotechnical problems, liquid water is relatively incompressible and if the voids are saturated with water, the pore water must be squeezed out in order to pack the particles closer together.

The principle of effective stress, introduced by Karl Terzaghi, states that the effective stress σ' (i.e., the average intergranular stress between solid particles) may be calculated by a simple subtraction of the pore pressure from the total stress:

$$\sigma' = \sigma - u$$

where σ is the total stress and u is the pore pressure. It is not practical to measure σ' directly, so in practice the vertical effective stress is calculated from the pore pressure and vertical total stress. The distinction between the terms pressure and stress is also important. By definition, pressure at a point is equal in all directions but stresses at a point can be different in different directions. In soil mechanics, compressive stresses and pressures are considered to be positive and tensile stresses are considered to be negative, which is different from the solid mechanics sign convention for stress.

Total Stress

For level ground conditions, the total vertical stress at a point, σ_v, on average, is the weight of everything above that point per unit area. The vertical stress beneath a uniform surface layer with density ρ, and thickness H is for example:

$$\sigma_v = \rho g H = \gamma H$$

where g is the acceleration due to gravity, and γ is the unit weight of the overlying layer. If there are multiple layers of soil or water above the point of interest, the vertical stress may be calculated by summing the product of the unit weight and thickness of all of the overlying layers. Total stress increases with increasing depth in proportion to the density of the overlying soil.

It is not possible to calculate the horizontal total stress in this way. Lateral earth pressures are addressed elsewhere.

Pore Water Pressure

Hydrostatic Conditions

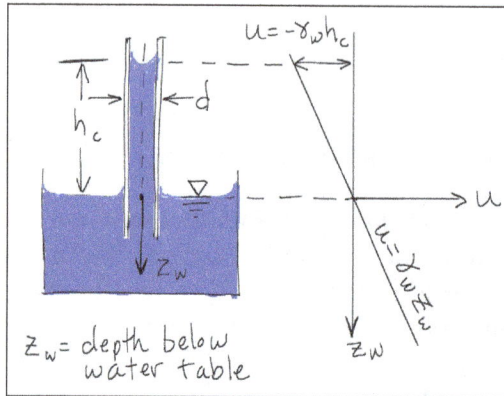

Water is drawn into a small tube by surface tension. Water pressure, u,
is negative above and positive below the free water surface.

If there is no pore water flow occurring in the soil, the pore water pressures will be hydrostatic. The water tableis located at the depth where the water pressure is equal to the atmospheric pressure. For hydrostatic conditions, the water pressure increases linearly with depth below the water table:

$$u = \rho_w g z_w$$

Where ρ_w is the density of water, and z_w is the depth above the water table.

Capillary Action

Due to surface tension, water will rise up in a small capillary tube above a free surface of water. Likewise, water will rise up above the water table into the small pore spaces around the soil particles. In fact the soil may be completely saturated for some distance above the water table. Above the height of capillary saturation, the soil may be wet but the water content will decrease with elevation. If the water in the capillary zone is not moving, the water pressure obeys the equation of hydrostatic equilibrium, $u = \rho_w g z_w$, but note that z_w, is negative above the water table. Hence, hydrostatic water pressures are negative above the water table. The thickness of the zone of capillary saturation depends on the pore size, but typically, the heights vary between a centimeter or so for coarse sand to tens of meters for a silt or clay. In fact the pore space of soil is a uniform fractal e.g. a set of uniformly distributed D-dimensional fractals of average linear size L. For the clay soil it has been found that L=0.15 mm and D=2.7.

The surface tension of water explains why the water does not drain out of a wet sand castle or a moist ball of clay. Negative water pressures make the water stick to the particles and pull the particles to each other, friction at the particle contacts make a sand castle stable. But as soon as a wet sand castle is submerged below a free water surface, the negative pressures are lost and the castle collapses. Considering the effective stress equation, $\sigma' = \sigma - u$, if the water pressure is negative, the effective stress may be positive, even on a free surface (a surface where the total normal stress is zero). The negative pore pressure pulls the particles together and causes compressive particle to particle contact forces. Negative pore pressures in clayey soil can be much more powerful than those in sand. Negative pore pressures explain why clay soils shrink when they dry and swell as they are wetted. The swelling and shrinkage can cause major distress, especially to light structures and roads.

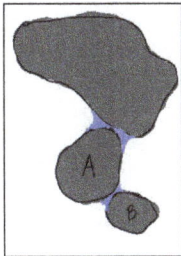

Water at particle contacts.   Intergranular contact force due to surface tension.   Shrinkage caused by drying.

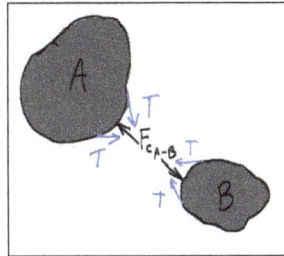

Consolidation: Transient Flow of Water

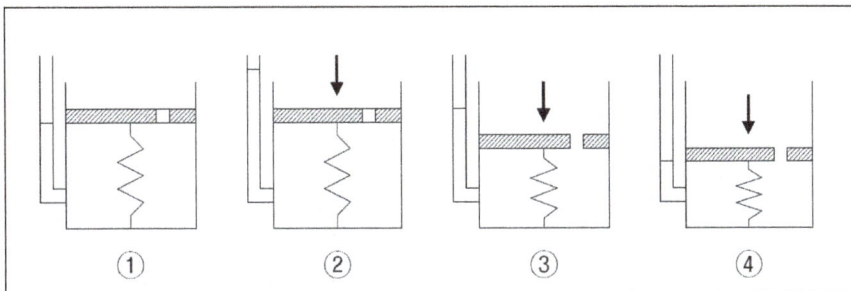

Consolidation analogy. The piston is supported by water underneath and a spring. When a load is applied to the piston, water pressure increases to support the load. As the water slowly leaks through the small hole, the load is transferred from the water pressure to the spring force.

Consolidation is a process by which soils decrease in volume. It occurs when stress is applied to a soil that causes the soil particles to pack together more tightly, therefore reducing volume. When this occurs in a soil that is saturated with water, water will be squeezed out of the soil. The time required to squeeze the water out of a thick deposit of clayey soil layer might be years. For a layer of sand, the water may be squeezed out in a matter of seconds. A building foundation or construction of a new embankment will cause the soil below to consolidate and this will cause settlement which in turn may cause distress to the building or embankment. Karl Terzaghi developed the theory of consolidation which enables prediction of the amount of settlement and the time required for the settlement to occur. Soils are tested with an oedometer test to determine their compression index and coefficient of consolidation.

When stress is removed from a consolidated soil, the soil will rebound, drawing water back into the pores and regaining some of the volume it had lost in the consolidation process. If the stress is reapplied, the soil will re-consolidate again along a recompression curve, defined by the re-compression index. Soil that has been consolidated to a large pressure and has been subsequently unloaded is considered to be *overconsolidated*. The maximum past vertical effective stress is termed the *preconsolidation stress*. A soil which is currently experiencing the maximum past vertical effective stress is said to be *normally consolidated*. The *overconsolidation ratio*, (OCR) is the ratio of the maximum past vertical effective stress to the current vertical effective stress. The OCR is significant for two reasons: firstly, because the compressibility of normally consolidated soil is significantly larger than that for overconsolidated soil, and secondly, the shear behavior and dilatancy of clayey soil are related to the OCR through critical state soil mechanics; highly overconsolidated clayey soils are dilatant, while normally consolidated soils tend to be contractive.

Shear Behavior: Stiffness and Strength

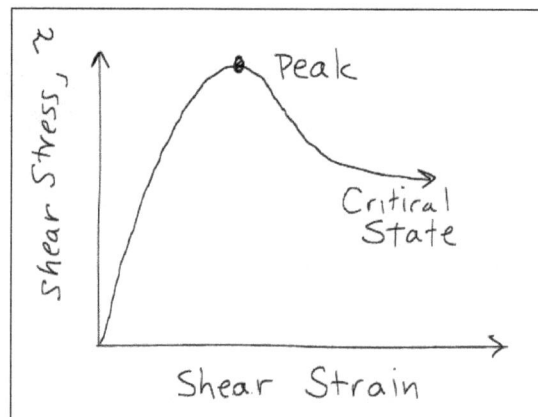

Typical stress strain curve for a drained dilatant soil.

The shear strength and stiffness of soil determines whether or not soil will be stable or how much it will deform. Knowledge of the strength is necessary to determine if a slope will be stable, if a building or bridge might settle too far into the ground, and the limiting pressures on a retaining wall. It is important to distinguish between failure of a soil element and the failure of a geotechnical structure (e.g., a building foundation, slope or retaining wall); some soil elements may reach their peak strength prior to failure of the structure. Different criteria can be used to define the "shear strength" and the "yield point" for a soil element from a stress–strain curve. One may define the peak shear strength as the peak of a stress–strain curve, or the shear strength at critical state as the value after large strains when the shear resistance levels off. If the stress–strain curve does not stabilize before the end of shear strength test, the "strength" is sometimes considered to be the shear resistance at 15–20% strain. The shear strength of soil depends on many factors including the effective stress and the void ratio.

The shear stiffness is important, for example, for evaluation of the magnitude of deformations of foundations and slopes prior to failure and because it is related to the shear wave velocity. The slope of the initial, nearly linear, portion of a plot of shear stress as a function of shear strain is called the shear modulus.

Friction, Interlocking and Dilation

Angle of repose.

Soil is an assemblage of particles that have little to no cementation while rock (such as sandstone) may consist of an assembly of particles that are strongly cemented together by chemical bonds. The shear strength of soil is primarily due to interparticle friction and therefore, the shear resistance on a plane is approximately proportional to the effective normal stress on that plane. The angle of internal friction is thus closely related to the maximum stable slope angle, often called the angle of repose.

But in addition to friction, soil derives significant shear resistance from interlocking of grains. If the grains are densely packed, the grains tend to spread apart from each other as they are subject to shear strain. The expansion of the particle matrix due to shearing was called dilatancy by Osborne Reynolds. If one considers the energy required to shear an assembly of particles there is energy input by the shear force, T, moving a distance, x and there is also energy input by the normal force, N, as the sample expands a distance, y. Due to the extra energy required for the particles to dilate against the confining pressures, dilatant soils have a greater peak strength than contractive soils. Furthermore, as dilative soil grains dilate, they become looser (their void ratio increases), and their rate of dilation decreases until they reach a critical void ratio. Contractive soils become denser as they shear, and their rate of contraction decreases until they reach a critical void ratio.

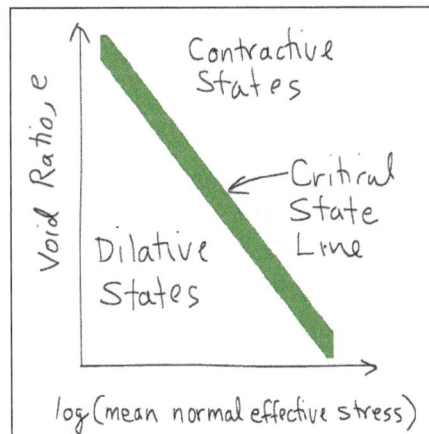

A critical state line separates the dilatant and contractive states for soil.

The tendency for a soil to dilate or contract depends primarily on the confining pressure and the void ratio of the soil. The rate of dilation is high if the confining pressure is small and the void ratio is small. The rate of contraction is high if the confining pressure is large and the void ratio is large. As a first approximation, the regions of contraction and dilation are separated by the critical state line.

Failure Criteria

After a soil reaches the critical state, it is no longer contracting or dilating and the shear stress on the failure plane τ_{crit} is determined by the effective normal stress on the failure plane $\sigma_{n'}$ and critical state friction angle ϕ_{crit}':

$$\tau_{crit} = \sigma_n' \tan \phi_{crit}' .$$

The peak strength of the soil may be greater, however, due to the interlocking (dilatancy) contribution. This may be stated:

$$\tau_{peak} = \sigma_n' \tan \phi_{peak}' .$$

Where $\phi_{peak}' > \phi_{crit}'$. However, use of a friction angle greater than the critical state value for design requires care. The peak strength will not be mobilized everywhere at the same time in a practical problem such as a foundation, slope or retaining wall. The critical state friction angle is not nearly as variable as the peak friction angle and hence it can be relied upon with confidence.

Not recognizing the significance of dilatancy, Coulomb proposed that the shear strength of soil may be expressed as a combination of adhesion and friction components:

$$\tau_f = c' + \sigma_f' \tan \phi' .$$

It is now known that the c' and ϕ' parameters in the last equation are not fundamental soil properties. In particular, c' and ϕ' are different depending on the magnitude of effective stress. According to Schofield, the longstanding use of c' in practice has led many engineers to wrongly believe that c' is a fundamental parameter. This assumption that c' and ϕ' are constant can lead to overestimation of peak strengths.

Structure, Fabric and Chemistry

In addition to the friction and interlocking (dilatancy) components of strength, the structure and fabric also play a significant role in the soil behavior. The structure and fabric include factors such as the spacing and arrangement of the solid particles or the amount and spatial distribution of pore water; in some cases cementitious material accumulates at particle-particle contacts. Mechanical behavior of soil is affected by the density of the particles and their structure or arrangement of the particles as well as the amount and spatial distribution of fluids present (e.g., water and air voids). Other factors include the electrical charge of the particles, chemistry of pore water, chemical bonds (i.e. cementation -particles connected through a solid substance such as recrystallized calcium carbonate).

Drained and Undrained Shear

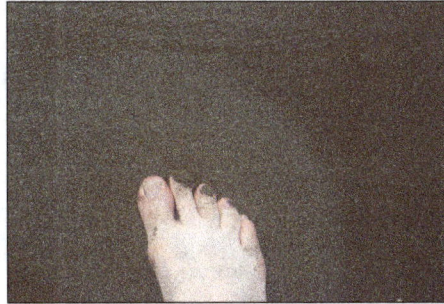

Foot pressure on sand causes it to dilate, drawing water from the surface into the pores.

The presence of nearly incompressible fluids such as water in the pore spaces affects the ability for the pores to dilate or contract.

If the pores are saturated with water, water must be sucked into the dilating pore spaces to fill the expanding pores (this phenomenon is visible at the beach when apparently dry spots form around feet that press into the wet sand).

Similarly, for contractive soil, water must be squeezed out of the pore spaces to allow contraction to take place.

Dilation of the voids causes negative water pressures that draw fluid into the pores, and contraction of the voids causes positive pore pressures to push the water out of the pores. If the rate of shearing is very large compared to the rate that water can be sucked into or squeezed out of the dilating or contracting pore spaces, then the shearing is called *undrained shear*, if the shearing is slow enough that the water pressures are negligible, the shearing is called *drained shear*. During undrained shear, the water pressure u changes depending on volume change tendencies. From the effective stress equation, the change in u directly effects the effective stress by the equation:

$$\sigma' = \sigma - u$$

The strength is very sensitive to the effective stress. It follows then that the undrained shear strength of a soil may be smaller or larger than the drained shear strength depending upon whether the soil is contractive or dilative.

Shear Tests

Strength parameters can be measured in the laboratory using direct shear test, triaxial shear test, simple shear test, fall cone test and (hand) shear vane test; there are numerous other devices and variations on these devices used in practice today. Tests conducted to characterize the strength and stiffness of the soils in the ground include the Cone penetration test and the Standard penetration test.

Other Factors

The stress–strain relationship of soils, and therefore the shearing strength, is affected by:

- Soil composition (basic soil material): Mineralogy, grain size and grain size distribution, shape of particles, pore fluid type and content, ions on grain and in pore fluid.

- State (initial): Define by the initial void ratio, effective normal stress and shear stress (stress history). State can be describe by terms such as: loose, dense, overconsolidated, normally consolidated, stiff, soft, contractive, dilative, etc.

- Structure: Refers to the arrangement of particles within the soil mass; the manner in which the particles are packed or distributed. Features such as layers, joints, fissures, slickensides, voids, pockets, cementation, etc., are part of the structure. Structure of soils is described by terms such as: undisturbed, disturbed, remolded, compacted, cemented; flocculent, honey-combed, single-grained; flocculated, deflocculated; stratified, layered, laminated; isotropic and anisotropic.

- Loading conditions: Effective stress path -drained, undrained, and type of loading -magnitude, rate (static, dynamic), and time history (monotonic, cyclic).

Applications

Lateral Earth Pressure

Lateral earth stress theory is used to estimate the amount of stress soil can exert perpendicular to gravity. This is the stress exerted on retaining walls. A lateral earth stress coefficient, K, is defined as the ratio of lateral (horizontal) effective stress to vertical effective stress for cohesionless soils ($K=\sigma'_h/\sigma'_v$). There are three coefficients: at-rest, active, and passive. At-rest stress is the lateral stress in the ground before any disturbance takes place. The active stress state is reached when a wall moves away from the soil under the influence of lateral stress, and results from shear failure due to reduction of lateral stress. The passive stress state is reached when a wall is pushed into the soil far enough to cause shear failure within the mass due to increase of lateral stress. There are many theories for estimating lateral earth stress; some are empirically based, and some are analytically derived.

Bearing Capacity

The bearing capacity of soil is the average contact stress between a foundation and the soil which will cause shear failure in the soil. Allowable bearing stress is the bearing capacity divided by a factor of safety. Sometimes, on soft soil sites, large settlements may occur under loaded foundations without actual shear failure occurring; in such cases, the allowable bearing stress is determined with regard to the maximum allowable settlement. It is important during construction and design stage of a project to evaluate the subgrade strength. The California Bearing Ratio (CBR) test is commonly used to determine the suitability of a soil as a subgrade for design and construction. The field Plate Load Test is commonly used to predict the deformations and failure characteristics of the soil/subgrade and modulus of subgrade reaction (ks). The Modulus of subgrade reaction (ks) is used in foundation design, soil-structure interaction studies and design of highway pavements.

Slope Stability

The field of slope stability encompasses the analysis of static and dynamic stability of slopes of earth and rock-fill dams, slopes of other types of embankments, excavated slopes, and natural slopes in soil and soft rock.

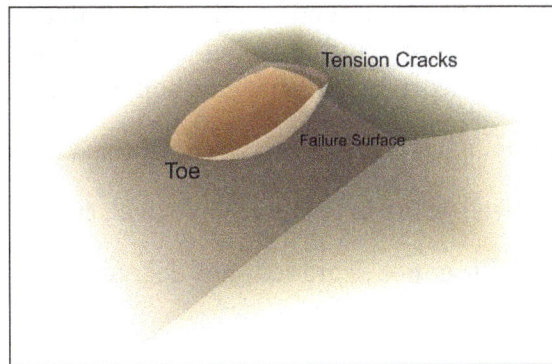

Simple slope slip section.

Earthen slopes can develop a cut-spherical weakness zone. The probability of this happening can be calculated in advance using a simple 2-D circular analysis package. A primary difficulty with analysis is locating the most-probable slip plane for any given situation. Many landslides have been analyzed only after the fact.

SOIL HEALTH

Soil health is a state of a soil meeting its range of ecosystem functions as appropriate to its environment. Soil health testing is an assessment of this status. Soil health depends on soil biodiversity (with a robust soil biota), and it can be improved via soil conditioning (soil amendment).

Aspects

The term soil health is used to describe the state of a soil in:

- Sustaining plant and animal productivity and biodiversity (Soil biodiversity);

- Maintaining or enhance water and air quality;

- Supporting human health and habitation.

Soil Health has partly if not largely replaced the expression "Soil Quality" that was extant in the 1990s. The primary difference between the two expressions is that soil quality was focused on individual traits within a functional group, as in "quality of soil for maize production" or "quality of soil for roadbed preparation" and so on. The addition of the word "health" shifted the perception to be integrative, holistic and systematic. The two expressions still overlap considerably.

The underlying principle in the use of the term "soil health" is that soil is not just an inert, lifeless growing medium, which modern farming tends to represent, rather it is a living, dynamic and ever-so-subtly changing whole environment. It turns out that soils highly fertile from the point of view of crop productivity are also lively from a biological point of view. It is now commonly recognized that soil microbial biomass is large: in temperate grassland soil the bacterial and fungal biomass have been documented to be 1–2 t (2.0 long tons; 2.2 short tons)/hectare and 2–5 t (4.9 long tons; 5.5 short tons)/ha, respectively. Some microbiologists now believe that 80% of soil nutrient functions are essentially controlled by microbes. If this is consistently true, than the prevailing Liebig nutrient theory model, and Mitscherlich modifications of it, all which exclude biology, are perhaps dangerously incorrect for managing soil fertility sustainably for the future.

Using the human health analogy, a healthy soil can be categorized as one:

- In a state of composite well-being in terms of biological, chemical and physical properties;

- Not diseased or infirmed (i.e. not degraded, nor degrading), nor causing negative off-site impacts;

- With each of its qualities cooperatively functioning such that the soil reaches its full potential and resists degradation;

- Providing a full range of functions (especially nutrient, carbon and water cycling) and in such a way that it maintains this capacity into the future.

Conceptualisation

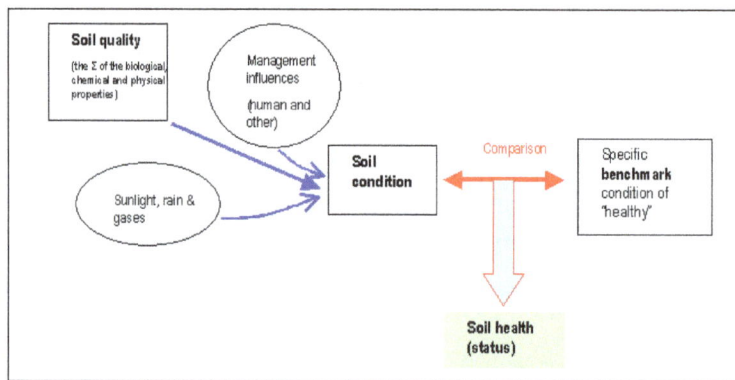

Soil health is the condition of the soil in a defined space and at a defined scale relative to a set of benchmarks that encompass healthy functioning. It would not be appropriate to refer to soil health for soil-roadbed preparation, as in the analogy of soil quality in a functional class. The definition of soil health may vary between users of the term as alternative users may place differing priorities upon the multiple functions of a soil. Therefore, the term soil health can only be understood within the context of the user of the term, and their aspirations of a soil, as well as by the boundary definition of the soil at issue.

Interpretation

Different soils will have different benchmarks of health depending on the "inherited" qualities, and on the geographic circumstance of the soil. The generic aspects defining a healthy soil can be considered as follows:

- "Productive" options are broad;

- Life diversity is broad;

- Absorbency, storing, recycling and processing is high in relation to limits set by climate;

- Water runoff quality is of high standard;

- Low entropy;

- No damage to, or loss of the fundamental components.

This translates to:

- A comprehensive cover of vegetation;

- Carbon levels relatively close to the limits set by soil type and climate;

- Little leakage of nutrients from the ecosystem;

- Biological and agricultural productivity relatively close to the limits set by the soil environment and climate;

- Only geological rates of erosion;

- No accumulation of contaminants;

- The ecosystem does not rely excessively on inputs of fossil energy.

Measurement

On the basis of the above, soil health will be measured in terms of individual ecosystem services provided relative to the benchmark. Specific benchmarks used to evaluate soil health include CO_2 release, humus levels, microbial activity, and available calcium.

Soil health testing is spreading in the United States, Australia and South Africa. Cornell University, a land-grant college in NY State, has had a Soil Health Test since 2006. Woods End Laboratories, a private soil lab founded in Maine in 1975, has offered a soil quality package since 1985 that contains physical (aggregate stability) chemical (mineral balance) and biology (CO_2 respiration) which today are prerequisites for soil health testing. in the United States, six commercial soil labs are registered for "Soil Health" testing under the NRCS website. The approach of all the labs is to add into common chemical nutrient testing a biological set of factors not normally included in routine soil testing. The best example is adding biological soil respiration ("CO_2-Burst") as a test procedure; this has already been adapted to modern commercial labs.

There is resistance among soil testing labs and university scientists to adding new biological tests, primarily since interpretation of soil fertility is based on models from "crop response" studies which match yield to test levels of specific chemical nutrients. These soil test methods have evolved slowly over the past 40 years and are backed by considerable effort. However, in this same time USA soils have also lost up to 75% of their carbon (humus), causing biological fertility to drop drastically. Many critics of the current system say this is sufficient evidence that the old soil testing models have failed us, and need to be replaced with new approaches. These older models have stressed "maximum yield" and " yield calibration" to such an extent that related factors have been overlooked. Thus, surface and groundwater pollution with excess nutrients (nitrates and phosphates) has grown enormously, and is reported presently (in the United States) to be the worst it has been since the 1970s, before the advent of environmental consciousness.

CROP ROTATION

Crop rotation is the practice of growing a series of dissimilar or different types of crops in the same area in sequenced seasons. It is done so that the soil of farms is not used for only one set of nutrients. It helps in reducing soil erosion and increases soil fertility and crop yield.

Growing the same crop in the same place for many years in a row (monocropping) gradually depletes the soil of certain nutrients. With rotation, a crop that leaches the soil of one kind of nutrient is followed during the next growing season by a dissimilar crop that returns that nutrient to the soil or draws a different ratio of nutrients. In addition, crop rotation mitigates the buildup of pathogens and pests that often occurs when one species is continuously cropped, and can also improve soil structure and fertility by increasing biomass from varied root structures.

Crop cycle is used in both conventional and organic farming systems.

Two-field System

Under a two-field rotation, half the land was planted in a year, while the other half lay fallow. Then, in the next year, the two fields were reversed. From the times of Charlemagne, farmers in Europe transitioned from a two-field crop rotation to a three-field crop rotation.

Three-field System

From the end of the Middle Ages until the 20th century, Europe's farmers practiced three-field rotation, dividing available lands into three parts. One section was planted in the autumn with rye or winter wheat, followed by spring oats or barley; the second section grew crops such as peas, lentils, or beans; and the third field was left fallow. The three fields were rotated in this manner so that every three years, a field would rest and be fallow. Under the two-field system, if one has a total of 600 acres (2.4 km²) of fertile land, one would only plant 300 acres. Under the new three-field rotation system, one would plant (and therefore harvest) 400 acres. But the additional crops had a more significant effect than mere quantitative productivity. Since the spring crops were mostly legumes, they increased the overall nutrition of the people of Northern Europe.

Four-field Rotation

Farmers in the region of Waasland (in present-day northern Belgium) pioneered a four-field rotation in the early 16th century, and the Britishagriculturist Charles Townshend popularised this system in the 18th century. The sequence of four crops (wheat, turnips, barley and clover), included a fodder crop and a grazing crop, allowing livestock to be bred year-round. The four-field crop rotation became a key development in the British Agricultural Revolution. The rotation between arable and ley is sometimes called ley farming.

Modern Developments

George Washington Carver studied crop-rotation methods in the United States, teaching southern farmers to rotate soil-depleting crops like cotton with soil-enriching crops like peanuts and peas.

In the Green Revolution of the mid-20th century the traditional practice of crop rotation gave way in some parts of the world to the practice of supplementing the chemical inputs to the soil through topdressing with fertilizers, adding (for example) ammonium nitrate or urea and restoring soil pHwith lime. Such practices aimed to increase yields, to prepare soil for specialist crops, and to reduce waste and inefficiency by simplifying planting and harvesting, irrigation .

Crop Choice

A preliminary assessment of crop interrelationships can be found in how each crop: (1) Contributes to soil organic matter (SOM) content, (2) Provides for pest management, (3) Manages deficient or excess nutrients, and (4) How it contributes to or controls for soil erosion.

Crop choice is often related to the goal the farmer is looking to achieve with the rotation, which could be weed management, increasing available nitrogen in the soil, controlling for erosion, or increasing soil structure and biomass, to name a few. When discussing crop rotations, crops are classified in different ways depending on what quality is being assessed: by family, by nutrient needs/benefits, and/or by profitability (i.e. cash crop versus cover crop). For example, giving adequate attention to plant family is essential to mitigating pests and pathogens. However, many farmers have success managing rotations by planning sequencing and cover crops around desirable cash crops. The following is a simplified classification based on crop quality and purpose.

Row Crops

Many crops which are critical for the market, like vegetables, are row crops (that is, grown in tight rows). While often the most profitable for farmers, these crops are more taxing on the soil. Row crops typically have low biomass and shallow roots: this means the plant contributes low residue to the surrounding soil and has limited effects on structure. With much of the soil around the plant exposed to disruption by rainfall and traffic, fields with row crops experience faster break down of organic matter by microbes, leaving fewer nutrients for future plants.

In short, while these crops may be profitable for the farm, they are nutrient depleting. Crop rotation practices exist to strike a balance between short-term profitability and long-term productivity.

Legumes

A great advantage of crop rotation comes from the interrelationship of nitrogen-fixing crops with nitrogen-demanding crops. Legumes, like alfalfa and clover, collect available nitrogen from the soil in nodules on their root structure. When the plant is harvested, the biomass of uncollected roots breaks down, making the stored nitrogen available to future crops. Legumes are also a valued green manure: a crop that collects nutrients and fixes them at soil depths accessible to future crops.

In addition, legumes have heavy tap roots that burrow deep into the ground, lifting soil for better tilth and absorption of water.

Grasses and Cereals

Cereal and grasses are frequent cover crops because of the many advantages they supply to soil quality and structure. The dense and far-reaching root systems give ample structure to surrounding soil and provide significant biomass for soil organic matter.

Grasses and cereals are key in weed management as they compete with undesired plants for soil space and nutrients.

Green Manure

Green manure is a crop that is mixed into the soil. Both nitrogen-fixing legumes and nutrient scavengers, like grasses, can be used as green manure. Green manure of legumes is an excellent source of nitrogen, especially for organic systems, however, legume biomass doesn't contribute to lasting soil organic matter like grasses do.

Planning a Rotation

There are numerous factors that must be taken into consideration when planning a crop rotation. Planning an effective rotation requires weighing fixed and fluctuating production circumstances: market, farm size, labor supply, climate, soil type, growing practices, etc. Moreover, a crop rotation must consider in what condition one crop will leave the soil for the succeeding crop and how one crop can be seeded with another crop. For example, a nitrogen-fixing crop, like a legume, should always precede a nitrogen depleting one; similarly, a low residue crop (i.e. a crop with low biomass) should be offset with a high biomass cover crop, like a mixture of grasses and legumes.

There is no limit to the number of crops that can be used in a rotation, or the amount of time a rotation takes to complete. Decisions about rotations are made years prior, seasons prior, or even at the very last minute when an opportunity to increase profits or soil quality presents itself. In short, there is no singular formula for rotation, but many considerations to take into account.

Implementation

Crop rotation systems may be enriched by the influences of other practices such as the addition of livestock and manure, intercropping or multiple cropping, and organic management low in pesticides and synthetic fertilizers.

Incorporation of Livestock

Introducing livestock makes the most efficient use of critical sod and cover crops; livestock (through manure) are able to distribute the nutrients in these crops throughout the soil rather than removing nutrients from the farm through the sale of hay.

In Sub-Saharan Africa, as animal husbandry becomes less of a nomadic practice many herders have begun integrating crop production into their practice. This is known as mixed farming, or the practice of crop cultivation with the incorporation of raising cattle, sheep and/or goats by the same economic entity, is increasingly common. This interaction between the animal, the land and the crops are being done on a small scale all across this region. Crop residues provide animal feed, while the animals provide manure for replenishing crop nutrients and draft power. Both processes are extremely important in this region of the world as it is expensive and logistically unfeasible to transport in synthetic fertilizers and large-scale machinery. As an additional benefit, the cattle, sheep and/or goat provide milk and can act as a cash crop in the times of economic hardship.

Organic Farming

Crop rotation is a required practice in order for a farm to receive organic certification in the United States. The "Crop Rotation Practice Standard" for the National Organic Program under the U.S. Code of Federal Regulations, states that:

> "Farmers are required to implement a crop rotation that maintains or builds soil organic matter, works to control pests, manages and conserves nutrients, and protects against erosion. Producers of perennial crops that aren't rotated may utilize other practices, such as cover crops, to maintain soil health".

In addition to lowering the need for inputs by controlling for pests and weeds and increasing available nutrients, crop rotation helps organic growers increase the amount of biodiversity on their farms. Biodiversity is also a requirement of organic certification, however, there are no rules in place to regulate or reinforce this standard. Increasing the biodiversity of crops has beneficial effects on the surrounding ecosystem and can host a greater diversity of fauna, insects, and beneficial microorganism in the soil. Some studies point to increased nutrient availability from crop rotation under organic systems compared to conventional practices as organic practices are less likely to inhibit of beneficial microbes in soil organic matter.

While multiple cropping and intercropping benefit from many of the same principals as crop rotation, they do not satisfy the requirement under the NOP.

Intercropping

Multiple cropping systems, such as intercropping or companion planting, offer more diversity and complexity within the same season or rotation, for example the three sisters. An example of companion planting is the inter-planting of corn with pole beans and vining squash or pumpkins. In this system, the beans provide nitrogen; the corn provides support for the beans and a "screen" against squash vine borer; the vining squash provides a weed suppressive canopy and a discouragement for corn-hungry raccoons.

Double-cropping is common where two crops, typically of different species, are grown sequentially in the same growing season, or where one crop (e.g. vegetable) is grown continuously with a cover crop (e.g. wheat). This is advantageous for small farms, who often cannot afford to leave cover crops to replenish the soil for extended periods of time, as larger farms can. When multiple cropping is implemented on small farms, these systems can maximize benefits of crop rotation on available land resources.

Benefits

Agronomists describe the benefits to yield in rotated crops as "The Rotation Effect". There are many found benefits of rotation systems: however, there is no specific scientific basis for the sometimes 10-25% yield increase in a crop grown in rotation versus monoculture. The factors related to the increase are simply described as alleviation of the negative factors of monoculture cropping systems. Explanations due to improved nutrition; pest, pathogen, and weed stress reduction; and improved soil structure have been found in some cases to be correlated, but causation has not been determined for the majority of cropping systems.

Other benefits of rotation cropping systems include production cost advantages. Overall financial risks are more widely distributed over more diverse production of crops and/or livestock. Less reliance is placed on purchased inputs and over time crops can maintain production goals with fewer inputs. This in tandem with greater short and long term yields makes rotation a powerful tool for improving agricultural systems.

Soil Organic Matter

The use of different species in rotation allows for increased soil organic matter (SOM), greater soil structure, and improvement of the chemical and biological soil environment for crops. With more SOM, water infiltration and retention improves, providing increased drought tolerance and decreased erosion.

Soil organic matter is a mix of decaying material from biomass with active microorganisms. Crop rotation, by nature, increases exposure to biomass from sod, green manure, and a various other plant debris. The reduced need for intensive tillage under crop rotation allows biomass aggregation to lead to greater nutrient retention and utilization, decreasing the need for added nutrients. With tillage, disruption and oxidation of soil creates a less conducive environment for diversity and proliferation of microorganisms in the soil. These microorganisms are what make nutrients available to plants. So, where "active" soil organic matter is a key to productive soil, soil with low microbial activity provides significantly fewer nutrients to plants; this is true even though the quantity of biomass left in the soil may be the same.

Soil microorganisms also decrease pathogen and pest activity through competition. In addition, plants produce root exudates and other chemicals which manipulate their soil environment as well as their weed environment. Thus rotation allows increased yields from nutrient availability but also alleviation of allelopathy and competitive weed environments.

Carbon Sequestration

Studies have shown that crop rotations greatly increase soil organic carbon (SOC) content, the main constituent of soil organic matter. Carbon, along with hydrogen and oxygen, is a macronutrient for plants. Highly diverse rotations spanning long periods of time have shown to be even more effective in increasing SOC, while soil disturbances (e.g. from tillage) are responsible for exponential decline in SOC levels. In Brazil, conversion to no-till methods combined with intensive crop rotations has been shown an SOC sequestration rate of 0.41 tonnes per hectare per year.

In addition to enhancing crop productivity, sequestration of atmospheric carbon has great implications in reducing rates of climate change by removing carbon dioxide from the air.

Nitrogen Fixing

Rotating crops adds nutrients to the soil. Legumes, plants of the family Fabaceae, for instance, have nodules on their roots which contain nitrogen-fixing bacteria called rhizobia. During a process called nodulation, the rhizobia bacteria use nutrients and water provided by the plant to convert atmospheric nitrogen into ammonia, which is then converted into an organic compound that the plant can use as its nitrogen source. It therefore makes good sense agriculturally to alternate them with cereals (family Poaceae) and other plants that require nitrates. How much nitrogen made available to the plants depends on factors such as the kind of legume, the effectiveness of rhizobia bacteria, soil conditions, and the availability of elements necessary for plant food.

Pathogen and Pest Control

Crop rotation is also used to control pests and diseases that can become established in the soil over time. The changing of crops in a sequence decreases the population level of pests by (1) Interrupting pest life cycles and (2) Interrupting pest habitat. Plants within the same taxonomic family tend to have similar pests and pathogens. By regularly changing crops and keeping the soil occupied by cover crops instead of lying fallow, pest cycles can be broken or limited, especially cycles that benefit from overwintering in residue. For example, root-knot nematode is a serious problem for some plants in warm climates and sandy soils, where it slowly builds up to high levels in the soil, and can severely damage plant productivity by cutting off circulation from the plant roots. Growing a crop that is not a host for root-knot nematode for one season greatly reduces the level of the nematode in the soil, thus making it possible to grow a susceptible crop the following season without needing soil fumigation.

This principle is of particular use in organic farming, where pest control must be achieved without synthetic pesticides.

Weed Management

Integrating certain crops, especially cover crops, into crop rotations is of particular value to weed management. These crops crowd out weed through competition. In addition, the sod and compost from cover crops and green manure slows the growth of what weeds are still able to make it through the soil, giving the crops further competitive advantage. By removing slowing the growth and proliferation of weeds while cover crops are cultivated, farmers greatly reduce the presence of weeds for future crops, including shallow rooted and row crops, which are less resistant to weeds. Cover crops are, therefore, considered conservation crops because they protect otherwise fallow land from becoming overrun with weeds.

This system has advantages over other common practices for weeds management, such as tillage. Tillage is meant to inhibit growth of weeds by overturning the soil; however, this has a countering effect of exposing weed seeds that may have gotten buried and burying valuable crop seeds. Under crop rotation, the number of viable seeds in the soil is reduced through the reduction of the weed population.

In addition to their negative impact on crop quality and yield, weeds can slow down the harvesting process. Weeds make farmers less efficient when harvesting, because weeds like bindweeds, and knotgrass, can become tangled in the equipment, resulting in a stop-and-go type of harvest.

Preventing Soil Erosion

Crop rotation can significantly reduce the amount of soil lost from erosion by water. In areas that are highly susceptible to erosion, farm management practices such as zero and reduced tillage can be supplemented with specific crop rotation methods to reduce raindrop impact, sediment detachment, sediment transport, surface runoff, and soil loss.

Protection against soil loss is maximized with rotation methods that leave the greatest mass of crop stubble (plant residue left after harvest) on top of the soil. Stubble cover in contact with the soil minimizes erosion from water by reducing overland flow velocity, stream power, and thus the ability of the water to detach and transport sediment. Soil Erosion and Cill prevent the disruption and detachment of soil aggregates that cause macropores to block, infiltration to decline, and runoff to increase. This significantly improves the resilience of soils when subjected to periods of erosion and stress.

When a forage crop breaks down, binding products are formed that act like an adhesive on the soil, which makes particles stick together, and form aggregates. The formation of soil aggregates is important for erosion control, as they are better able to resist raindrop impact, and water erosion. Soil aggregates also reduce wind erosion, because they are larger particles, and are more resistant to abrasion through tillage practices.

The effect of crop rotation on erosion control varies by climate. In regions under relatively consistent climate conditions, where annual rainfall and temperature levels are assumed, rigid crop rotations can produce sufficient plant growth and soil cover. In regions where climate conditions are less predictable, and unexpected periods of rain and drought may occur, a more flexible approach for soil cover by crop rotation is necessary. An opportunity cropping system promotes adequate soil cover under these erratic climate conditions. In an opportunity cropping system, crops are grown when soil water is adequate and there is a reliable sowing window. This form of cropping system is likely to produce better soil cover than a rigid crop rotation because crops are only sown under optimal conditions, whereas rigid systems are not necessarily sown in the best conditions available.

Crop rotations also affect the timing and length of when a field is subject to fallow. This is very important because depending on a particular region's climate, a field could be the most vulnerable to erosion when it is under fallow. Efficient fallow management is an essential part of reducing erosion in a crop rotation system. Zero tillage is a fundamental management practice that promotes crop stubble retention under longer unplanned fallows when crops cannot be planted. Such management practices that succeed in retaining suitable soil cover in areas under fallow will ultimately reduce soil loss. In a recent study that lasted a decade, it was found that a common winter cover crop after potato harvest such as fall rye can reduce soil run-off by as much as 43%, and this is typically the most nutritional soil.

Biodiversity

Increasing the biodiversity of crops has beneficial effects on the surrounding ecosystem and can host a greater diversity of fauna, insects, and beneficial microorganisms in the soil. Some studies point to increased nutrient availability from crop rotation under organic systems compared to conventional practices as organic practices are less likely to inhibit of beneficial microbes in soil organic matter, such as arbuscular mycorrhizae, which increase nutrient uptake in plants. Increasing biodiversity also increases the resilience of agro-ecological systems.

Farm Productivity

Crop rotation contributes to increased yields through improved soil nutrition. By requiring planting and harvesting of different crops at different times, more land can be farmed with the same amount of machinery and labour.

Challenges

Different crops in the rotation can reduce the risks of adverse weather for the individual farmer.While crop rotation requires a great deal of planning, crop choice must respond to a number of fixed conditions (soil type, topography, climate, and irrigation) in addition to conditions that may change dramatically from year to the next (weather, market, labor supply). In this way, it is unwise to plan crops years in advance. Improper implementation of a crop rotation plan may lead to imbalances in the soil nutrient composition or a buildup of pathogens affecting a critical crop. The consequences of faulty rotation may take years to become apparent even to experienced soil scientists and can take just as long to correct.

Many challenges exist within the practices associated with crop rotation. For example, green manure from legumes can lead to an invasion of snails or slugs and the decay from green manure can occasionally suppress the growth of other crops.

NUTRIENT MANAGEMENT

Nitrogen fertilizer being applied to growing corn (maize) in a contoured, no-tilled field in Iowa.

Nutrient management is the science and practice directed to link soil, crop, weather, and hydrologic factors with cultural, irrigation, and soil and water conservation practices to achieve optimal nutrient use efficiency, crop yields, crop quality, and economic returns, while reducing off-site transport of nutrients (fertilizer) that may impact the environment. It involves matching a specific field soil, climate, and crop management conditions to rate, source, timing, and place (commonly known as the 4R nutrient stewardship) of nutrient application.

Important factors that need to be considered when managing nutrients include (a) the application of nutrients considering the achievable optimum yields and, in some cases, crop quality; (b) the management, application, and timing of nutrients using a budget based on all sources and sinks active at the site; and (c) the management of soil, water, and crop to minimize the off-site transport of nutrients from nutrient leaching out of the root zone, surface runoff, and volatilization (or other gas exchanges).

There can be potential interactions because of differences in nutrient pathways and dynamics. For instance, practices that reduce the off-site surface transport of a given nutrient may increase the leaching losses of other nutrients. These complex dynamics present nutrient managers the difficult task of achieve the best balance for maximizing profit while contributing to the conservation of our biosphere.

Nutrient Management Plan

Manure spreader.

A crop nutrient management plan is a tool that farmers can use to increase the efficiency of all the nutrient sources a crop uses while reducing production and environmental risk, ultimately increasing profit. It is generally agreed that there are ten fundamental components of a Crop Nutrient Management Plan. Each component is critical to helping analyze each field and improve nutrient efficiency for the crops grown. These components include:

Field Map

The map, including general reference points (such as streams, residences, wellheads etc.), number of acres, and soil types is the base for the rest of the plan.

Soil Test

How much of each nutrient (N-P-K and other critical elements such as pH and organic matter) is in the soil profile? The soil test is a key component needed for developing the nutrient rate recommendation.

Crop Sequence

Did the crop that grew in the field last year (and in many cases two or more years ago) fix nitrogen for use in the following years? Has long-term no-till increased organic matter? Did the end-of-season stalk test show a nutrient deficiency? These factors also need to be factored into the plan.

Estimated Yield

Factors that affect yield are numerous and complex. A field's soils, drainage, insect, weed and crop disease pressure, rotation and many other factors differentiate one field from another. This is why using historic yields is important in developing yield estimates for next year. Accurate yield estimates can improve nutrient use efficiency.

Sources and Forms

The sources and forms of available nutrients can vary from farm-to-farm and even field-to-field. For instance, manure fertility analysis, storage practices and other factors will need to be included in a nutrient management plan. Manure nutrient tests/analysis are one way to determine the fertility of it. Nitrogen fixed from a previous year's legume crop and residual affects of manure also effects rate recommendations. Many other nutrient sources should also be factored into this plan.

Sensitive Areas

What's out of the ordinary about a field's plan? Is it irrigated? Next to a stream or lake? Especially sandy in one area? Steep slope or low area? Manure applied in one area for generations due to proximity of dairy barn? Extremely productive—or unproductive—in a portion of the field? Are there buffers that protect streams, drainage ditches, wellheads, and other water collection points? How far away are the neighbors? What's the general wind direction? This is the place to note these and other special conditions that need to be considered.

Recommended Rates

Here's the place where science, technology, and art meet. Given everything you've noted, what is the optimum rate of N, P, K, lime and any other nutrients? While science tells us that a crop has changing nutrient requirements during the growing season, a combination of technology and farmer's management skills assure nutrient availability at all stages of growth. No-till corn generally requires starter fertilizer to give the seedling a healthy start.

Recommended Timing

When does the soil temperature drop below 50 degrees? Will a N stabilizer be used? What's the tillage practice? Strip-till corn and no-till often require different timing approaches than seed planted into a field that's been tilled once with a field cultivator. Will a starter fertilizer be used to give the seedling a healthy start? How many acres can be covered with available labor (custom or hired) and equipment? Does manure application in a farm depend on a custom applicator's schedule? What agreements have been worked out with neighbors for manure use on their fields? Is a neighbor hosting a special event? All these factors and more will likely figure into the recommended timing.

Recommended Methods

Surface or injected? While injection is clearly preferred, there may be situations where injection is not feasible (i.e. pasture, grassland). Slope, rainfall patterns, soil type, crop rotation and many other factors determine which method is best for optimizing nutrient efficiency (availability and loss) in farms. The combination that's right in one field may differ in another field even with the same crop.

Annual Review and Update

Even the best managers are forced to deviate from their plans. What rate was actually applied? Where? Using which method? Did an unusually mild winter or wet spring reduce soil nitrate? Did a dry summer, disease, or some other unusual factor increase nutrient carryover? These and other factors should be noted as they occur.

When such a plan is designed for animal feeding operations (AFO), it may be termed a "manure management plan." In the United States, some regulatory agencies recommend or require that farms implement these plans in order to prevent water pollution. The U.S. Natural Resources Conservation Service (NRCS) has published guidance documents on preparing a comprehensive nutrient management plan (CNMP) for AFOs.

The International Plant Nutrition Institute has published a 4R plant nutrition manual for improving the management of plant nutrition. The manual outlines the scientific principles behind each of the four R's or "rights" (right source of nutrient, right application rate, right time, right place) and discusses the adoption of 4R practices on the farm, approaches to nutrient management planning, and measurement of sustainability performance.

Nitrogen Management

Of the 16 essential plant nutrients, nitrogen is usually the most difficult to manage in field crop systems. This is because the quantity of plant-available nitrogen can change rapidly in response to changes in soil water status. Nitrogen can be lost from the plant-soil system by one or more of the following processes: leaching; surface runoff; soil erosion; ammonia volatilization; and denitrification.

Nitrogen Management Practices that Improve Nitrogen Efficiency

Nitrogen management aims to maximize the efficiency with which crops use applied N. Improvements in N use efficiency are associated with decreases in N loss from the soil. Although losses cannot be avoided completely, significant improvements can be realized by applying one or more of the following management practices in the cropping system.

Reduction of Greenhouse Gas Emissions

- Climate Smart Agriculture includes the use of 4R Nutrient Stewardship principles to reduce field emissions of nitrous oxide (N_2O) from the application of nitrogen fertilizer. Nitrogen fertilizer is an important driver of nitrous oxide emissions, but it is also the main driver of yield in modern high production systems. Through careful selection of nitrogen fertilizer source, rate, timing and placement practices, the nitrous oxide emissions per unit of crop produced can be substantially reduced, in some cases by up to half. The practices that reduce nitrous oxide emissions also tend to increase nitrogen use efficiency and the economic return on fertilizer dollars.

Reduction of N Loss in Runoff Water and Eroded Soil

- No-till, conservation tillage and other runoff control measures reduce N loss in surface runoff and eroded soil material.

- The use of daily estimates of soil moisture and crop needs to schedule irrigation reduces the risk of surface runoff and soil erosion.

Reduction of the Volatilization of N as Ammonia Gas

- Incorporation and/or injection of urea and ammonium-containing fertilizers decreases ammonia volatilization because good soil contact buffers pH and slows the generation of ammonia gas from ammonium ions.

- Urease inhibitors temporarily block the function of the urease enzyme, maintaining urea-based fertilizers in the non-volatile urea form, reducing volatilization losses when these fertilizers are surface applied; these losses can be meaningful in high-residue, conservation tillage systems.

Prevention of the Build-up of High Soil Nitrate Concentrations

Nitrate is the form of nitrogen that is most susceptible to loss from the soil, through denitrification and leaching. The amount of N lost via these processes can be limited by restricting soil nitrate concentrations, especially at times of high risk. This can be done in many ways, although these are not always cost-effective.

Nitrogen Rates

Rates of N application should be high enough to maximize profits in the long term and minimize residual (unused) nitrate in the soil after harvest.

- The use of local research to determine recommended nitrogen application rates should result in appropriate N rates.

- Recommended N application rates often rely on an assessment of yield expectations - these should be realistic, and preferably based on accurate yield records.

- Fertilizer N rates should be corrected for N that is likely to be mineralized from soil organic matter and crop residues (especially legume residues).

- Fertilizer N rates should allow for N applied in manure, in irrigation water, and from atmospheric deposition.

- Where feasible, appropriate soil tests can be used to determine residual soil N.

Soil Testing for N

- Preplant soil tests provide information on the soil's N-supply power.

- Late spring or pre-side-dress N tests can determine if and how much additional N is needed.

- New soil test and sampling procedures, such as amino sugar tests, grid mapping, and real-time sensors can refine N requirements.

- Post-harvest soil tests determine if N management the previous season was appropriate.

Crop Testing for N

- Plant tissue tests can identify N deficiencies.

- Sensing variations in plant chlorophyll content facilitates variable rate N applications in-season.

- Post-black-layer corn stalk nitrate tests help to determine if N rates were low, optimal, or excessive in the previous crop, so that management changes can be made in following crops.

Precision Agriculture

- Variable rate applicators, combined with intensive soil or crop sampling, allow more precise and responsive application rates.

Timing of N Applications

- Apply N close to the time when crops can utilize it.

- Make side-dress N applications close to the time of most rapid N uptake.

- Split applications, involving more than one application, allow efficient use of applied N and reduce the risk of N loss to the environment.

N Forms, Including Slow or Controlled Release Fertilizers and Inhibitors

- Slow or controlled release fertilizer delays the availability of nitrogen to the plant until a time that is more appropriate for plant uptake - the risk of N loss through denitrification and leaching is reduced by limiting nitrate concentrations in the soil.

- Nitrification inhibitors maintain applied N in the ammonium form for a longer period of time, thereby reducing leaching and denitrification losses.

N capture

- Particular crop varieties are able to more efficiently extract N from the soil and improve N use efficiency. Breeding of crops for efficient N uptake is in progress.

- Rotation with deep-rooted crops helps capture nitrates deeper in the soil profile.

- Cover crops capture residual nitrogen after crop harvest and recycle it as plant biomass.

- Elimination of restrictions to subsoil root development; subsoil compaction and subsoil acidity prevent root penetration in many subsoils worldwide, promoting build-up of subsoil nitrate concentrations which are susceptible to denitrification and leaching when conditions are suitable.

- Good agronomic practice, including appropriate plant populations and spacing and good weed and pest management, allows crops to produce large root systems to optimse N capture and crop yield.

Water Management

Conservation Tillage

Conservation tillage optimizes soil moisture conditions that improve water use efficiency; in water-stressed conditions, this improves crop yield per unit N applied.

N Fertilizer Application Method and Placement

- In ridged crops, placing N fertilizers in a band in ridges makes N less susceptible to leaching.

- Row fertilizer applicators, such as injectors, which form a compacted soil layer and surface ridge, can reduce N losses by diverting water flow.

Good Irrigation Management can Dramatically Improve N-use Efficiency

- Scheduled irrigation based on soil moisture estimates and daily crop needs will improve both water-use and N-use efficiency.

- Sprinkler irrigation systems apply water more uniformly and in lower amounts than furrow or basin irrigation systems.

- Furrow irrigation efficiency can be improved by adjusting set time, stream size, furrow length, watering every other row, or the use of surge valves.

- Alternate row irrigation and fertilization minimizes water contact with nutrients.

- Application of N fertilizer through irrigation systems (fertigation) facilitates N supply when crop demand is greatest.

- Polyacrylamide (PAM) treatment during furrow irrigation reduces sediment and N losses.

Drainage Systems

- Some subirrigation systems recycle nitrate leached from the soil profile and reduce nitrate lost in drainage water.

- Excessive drainage can lead to rapid through-flow of water and N leaching, but restricted or insufficient drainage favors anaerobic conditions and denitrification.

Use of Simulation Models

Short-term changes in the plant-available N status make accurate seasonal predictions of crop N requirement difficult in most situations. However, models (such as NLEAP and Adapt-N) that use soil, weather, crop, and field management data can be updated with day-to-day changes and thereby improve predictions of the fate of applied N. They allows farmers to make adaptive management decisions that can improve N-use efficiency and minimize N losses and environmental impact while maximizing profitability.

Additional Measures to Minimize Environmental Impact

Conservation Buffers

- Buffers trap sediment containing ammonia and organic N.

- Nitrate in subsurface flow is reduced through denitrification enhanced by carbon energy sources contained in the soil associated with buffer vegetation.

- Buffer vegetation takes up nitrogen, other nutrients, and reduces loss to water.

Constructed Wetlands

Constructed wetlands located strategically on the landscape to process drainage effluent reduces sediment and nitrate loads to surface water.

SOIL CONDITIONING

Soil conditioning means improving several aspects of the quality of soil:

- Tilth: This refers to the soil's physical condition and larger-scale structure. It includes whether the soil has aggregates (clumps) and what size they are, whether it has channels where water can enter and drain, and its level of aeration. Soil with good tilth has a structure that supports healthy root growth.

- Water holding capacity: This is partially a function of the soil type, but there are other things that alter it. Ideally, soil is well drained but holds enough water to support healthy plant growth.

- Nutrient holding capacity: This refers to the soil's ability to hold onto minerals that plants use as nutrients. Clay soils typically have greater nutrient holding capacity, which means they have the potential to be very fertile. However, they may need work to overcome some other disadvantages, like their tendency to become compacted or clumpy.

- Percentage of organic matter: This is very important in promoting soil biological activity, and it affects the water and nutrient holding capacity and the tilth.

Ways to Condition Soil

First, avoid degrading soil quality. Walking on garden soil, allowing bare ground to be exposed to rainfall or flooding, and working soil when it's too wet can all harm tilth. In soil that is low in organic matter, over-working soil can cause a hard crust to form. Exposing bare soil to the elements can also worsen quality, so keep soil covered between crops, such as with tarps, mulch, or cover crops.

Then, think about what changes your soil needs and how you can achieve them. Using soil conditioners (amendments that are meant to improve soil physical condition) is one way to do this.

Adding organic matter in the form of compost, manure, or readily available materials like coffee grounds is a reliable means of improving soil quality. These soil conditioners both improve the water retention of sandy soils and improve the drainage of clay soils that tend to become waterlogged. It is usually easier to maintain good tilth in soil that is high in organic matter. And compost provides long-lasting benefits by increasing soil nutrient content and contributing to the biological activity of soil.

Other Methods for Conditioning Soil

Compost is good for almost any soil. But some soil conditioners, such as gypsum and peat, provide benefits only for certain soil types or certain types of plants.

Other products sold as soil conditioners have dubious benefits, or the benefits are unknown. Before using soil conditioners, check for reliable evidence of the product's effectiveness. Some would need to be added in impractically large amounts to change the properties of your soil.

Planting cover crops can help you protect bare ground and add organic matter in addition to improving tilth. Taproot crops like forage radish, alfalfa, and chicory can help form channels that allow water to move through compacted or poorly drained soils.

NUTRIENT BUDGETING

Nutrient budgets offer insight into the balance between crop inputs and outputs. In short, they compare nutrients applied to the soil to nutrients taken up by crops. A nutrient budget takes into account all the nutrient inputs on a farm and all those removed from the land. The most obvious source of nutrients in this situation is fertilizer, but this is only part of the picture. Other inputs come with rainfall, in supplements brought on to the farm and in effluent – either farm or dairy factory – spread on the land. In addition, nutrients can be moved around the farm – from an area used for growing silage to the area used to feed it out, from paddock to raceway, and within paddocks in dung and urine patches. Nutrients are removed from the farm in stock sold on, products (meat, milk, wool), crops sold or fed out off farm, and through processes such as nitrate leaching, volatilization and phosphate run-off etc.

Importance

An accurate nutrient budget is an important tool to provide an early indication of potential problems arising from (i) A nutrient surplus, leading to an accumulation of nutrients and increased risk of loss or (ii) A deficit, depleting nutrient reserves and increasing the risk of deficiencies and reduced crop yields. They also provide regulatory authorities with a readily-determined, comparative indicator of environmental impact. Overall, nutrient budgets help ensure that farming practices are conducted in an efficient, economic, and environmentally sustainable manner.

Contents

A nutrient budget isn't as exact as a financial statement. An assortment of variables affects each tract of land. For example, some areas may have had too much manure applied over time or it may have been unevenly distributed, and previous flooding could affect results. Limits and assumptions should be incorporated when compiling a budget including the average nutrient removal coefficient values if they are not specific to a certain field.

Soil Test

This component is complementary to the budget and lets you know what nutrients are already available to crops and helps you plan input purchases. It is a critical best management practice (BMP) in the 4R strategy.

Yield History

By examining the historical yields of crops take from specific fields, you can calculate nutrient removal over time. Yield history may also help better predict the amount of uptake that will occur with similar crops planted in the future.

Previous Applications

Knowing what's been applied to the field in years past will offer insight into what may already be in the ground or what nutrients may no longer be present.

Water

Consider what kind of water has been applied to the field. Does irrigation water contain dissolved nutrients such as nitrogen (N), sulfur (S), or chloride (Cl)? If so, it should be counted as input.

Look Around

Consider water sources that could run into your field. Is there a manufacturing facility nearby? What makes up these water sources can impact how you plant.

COVER CROP

A cover crop is planted to manage soil erosion, soil fertility, soil quality, water, weeds, pests, diseases, biodiversity and wildlife in an *agroecosystem*—an ecological system managed and shaped by humans. Cover crops may be an off-season crop planted after harvesting the cash crop. The cover crop may grow over winter.

Soil Erosion

Although cover crops can perform multiple functions in an agroecosystem simultaneously, they are often grown for the sole purpose of preventing soil erosion. Soil erosion is a process that can irreparably reduce the productive capacity of an agroecosystem. Dense cover crop stands physically slow down the velocity of rainfall before it contacts the soil surface, preventing soil splashing and erosive surface runoff. Additionally, vast cover crop root networks help anchor the soil in place and increase soil porosity, creating suitable habitat networks for soil macrofauna. It keeps the enrichment of the soil good for the next few years.

Soil Fertility Management

One of the primary uses of cover crops is to increase soil fertility. These types of cover crops are referred to as "green manure." They are used to manage a range of soil macronutrients and micronutrients. Of the various nutrients, the impact that cover crops have on nitrogen management has received the most attention from researchers and farmers, because nitrogen is often the most limiting nutrient in crop production.

Often, green manure crops are grown for a specific period, and then plowed under before reaching full maturity in order to improve soil fertility and quality. Also the stalks left block the soil from being eroded.

Green manure crops are commonly leguminous, meaning they are part of the pea family, Fabaceae. This family is unique in that all of the species in it set pods, such as bean, lentil, lupins and alfalfa. Leguminous cover crops are typically high in nitrogen and can often provide the required quantity of nitrogen for crop production. In conventional farming, this nitrogen is typically applied in chemical fertilizer form. This quality of cover crops is called fertilizer replacement value.

Another quality unique to leguminous cover crops is that they form symbiotic relationships with the rhizobial bacteria that reside in legume root nodules. Lupins is nodulated by the soil microorganism *Bradyrhizobium* sp. (Lupinus). Bradyrhizobia are encountered as microsymbionts in other leguminous crops (*Argyrolobium, Lotus, Ornithopus, Acacia, Lupinus*) of Mediterranean origin. These bacteria convert biologically unavailable atmospheric nitrogen gas (N_2) to biologically available ammonium (NH^+_4) through the process of biological nitrogen fixation.

Prior to the advent of the Haber-Bosch process, an energy-intensive method developed to carry out industrial nitrogen fixation and create chemical nitrogen fertilizer, most nitrogen introduced to ecosystems arose through biological nitrogen fixation. Some scientists believe that widespread biological nitrogen fixation, achieved mainly through the use of cover crops, is the only alternative to industrial nitrogen fixation in the effort to maintain or increase future food production levels. Industrial nitrogen fixation has been criticized as an unsustainable source of nitrogen for food production due to its reliance on fossil fuel energy and the environmental impacts associated with chemical nitrogen fertilizer use in agriculture. Such widespread environmental impacts include nitrogen fertilizer losses into waterways, which can lead to eutrophication (nutrient loading) and ensuing hypoxia (oxygen depletion) of large bodies of water.

An example of this lies in the Mississippi Valley Basin, where years of fertilizer nitrogen loading into the watershed from agricultural production have resulted in a hypoxic "dead zone" off the Gulf of Mexico the size of New Jersey. The ecological complexity of marine life in this zone has been diminishing as a consequence.

As well as bringing nitrogen into agro ecosystems through biological nitrogen fixation, types of cover crops known as "catch crops" are used to retain and recycle soil nitrogen already present. The catch crops take up surplus nitrogen remaining from fertilization of the previous crop, preventing it from being lost through leaching, or gaseous denitrification or volatilization.

Catch crops are typically fast-growing annual cereal species adapted to scavenge available nitrogen efficiently from the soil. The nitrogen tied up in catch crop biomass is released back into the soil once the catch crop is incorporated as a green manure or otherwise begins to decompose.

An example of green manure use comes from Nigeria, where the cover crop *Mucuna pruriens* (velvet bean) has been found to increase the availability of phosphorus in soil after a farmer applies rock phosphate.

Soil Quality Management

Cover crops can also improve soil quality by increasing soil organic matter levels through the input of cover crop biomass over time. Increased soil organic matter enhances soil structure, as well as the water and nutrient holding and buffering capacity of soil. It can also lead to increased soil carbon sequestration, which has been promoted as a strategy to help offset the rise in atmospheric carbon dioxide levels.

Soil quality is managed to produce optimum circumstances for crops to flourish. The principal factors of soil quality are soil salination, pH, microorganism balance and the prevention of soil contamination.

Water Management

By reducing soil erosion, cover crops often also reduce both the rate and quantity of water that drains off the field, which would normally pose environmental risks to waterways and ecosystems downstream. Cover crop biomass acts as a physical barrier between rainfall and the soil surface, allowing raindrops to steadily trickle down through the soil profile. Also, as stated above, cover crop root growth results in the formation of soil pores, which in addition to enhancing soil macrofauna habitat provides pathways for water to filter through the soil profile rather than draining off the field as surface flow. With increased water infiltration, the potential for soil water storage and the recharging of aquifers can be improved.

Just before cover crops are killed (by such practices including mowing, tilling, discing, rolling, or herbicide application) they contain a large amount of moisture. When the cover crop is incorporated into the soil, or left on the soil surface, it often increases soil moisture. In agro ecosystems where water for crop production is in short supply, cover crops can be used as a mulch to conserve water by shading and cooling the soil surface. This reduces evaporation of soil moisture. In other situations farmers try to dry the soil out as quickly as possible going into the planting season. Here prolonged soil moisture conservation can be problematic.

While cover crops can help to conserve water, in temperate regions (particularly in years with below average precipitation) they can draw down soil water supply in the spring, particularly if climatic growing conditions are good. In these cases, just before crop planting, farmers often face a trade-off between the benefits of increased cover crop growth and the drawbacks of reduced soil moisture for cash crop production that season. C/N ratio is balanced with this application.

Weed Management

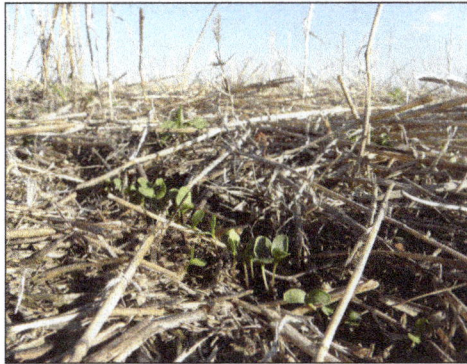

Cover crop in South Dakota.

Thick cover crop stands often compete well with weeds during the cover crop growth period, and can prevent most germinated weed seeds from completing their life cycle and reproducing. If the cover crop is flattened down on the soil surface rather than incorporated into the soil as a green manure after its growth is terminated, it can form a nearly impenetrable mat. This drastically reduces light transmittance to weed seeds, which in many cases reduces weed seed germination rates. Furthermore, even when weed seeds germinate, they often run out of stored energy for growth before building the necessary structural capacity to break through the cover crop mulch layer. This is often termed the cover crop smother effect.

Some cover crops suppress weeds both during growth and after death. During growth these cover crops compete vigorously with weeds for available space, light, and nutrients, and after death they smother the next flush of weeds by forming a mulch layer on the soil surface. For example, researchers found that when using *Melilotus officinalis* (yellow sweet clover) as a cover crop in an improved fallow system (where a fallow period is intentionally improved by any number of different management practices, including the planting of cover crops), weed biomass only constituted between 1-12% of total standing biomass at the end of the cover crop growing season. Furthermore, after cover crop termination, the yellow sweet clover residues suppressed weeds to levels 75–97% lower than in fallow (no yellow sweet clover) systems.

In addition to competition-based or physical weed suppression, certain cover crops are known to suppress weeds through allelopathy. This occurs when certain biochemical cover crop compounds are degraded that happen to be toxic to, or inhibit seed germination of, other plant species. Some well known examples of allelopathic cover crops are *Secale cereale* (rye), *Vicia villosa* (hairy vetch), *Trifolium pratense* (red clover), *Sorghum bicolor* (sorghum-sudangrass), and species in the family Brassicaceae, particularly mustards. In one study, rye cover crop residues were found to have provided between 80% and 95% control of early season broadleaf weeds when used as a mulch during the production of different cash crops such as soybean, tobacco, corn, and sunflower.

In a recent study released by the Agricultural Research Service (ARS) scientists examined how rye seeding rates and planting patterns affected cover crop production. The results show that planting more pounds per acre of rye increased the cover crop's production as well as decreased the amount of weeds. The same was true when scientists tested seeding rates on legumes and oats; a higher density of seeds planted per acre decreased the amount of weeds and increased the yield of legume and oat production. The planting patterns, which consisted of either traditional rows or grid patterns, did not seem to make a significant impact on the cover crop's production or on the weed production in either cover crop. The ARS scientists concluded that increased seeding rates could be an effective method of weed control.

Disease Management

In the same way that allelopathic properties of cover crops can suppress weeds, they can also break disease cycles and reduce populations of bacterial and fungal diseases, and parasitic nematodes. Species in the family Brassicaceae, such as mustards, have been widely shown to suppress fungal disease populations through the release of naturally occurring toxic chemicals during the degradation of glucosinolade compounds in their plant cell tissues.

Pest Management

Some cover crops are used as so-called "trap crops", to attract pests away from the crop of value and toward what the pest sees as a more favorable habitat. Trap crop areas can be established within crops, within farms, or within landscapes. In many cases the trap crop is grown during the same season as the food crop being produced. The limited area occupied by these trap crops can be treated with a pesticide once pests are drawn to the trap in large enough numbers to reduce the pest populations. In some organic systems, farmers drive over the trap crop with a large vacuum-based implement to physically pull the pests off the plants and out of the field. This system has been recommended for use to help control the lygus bugsin organic strawberry production. Another example of trap crops are nematode resistance white mustard (*Sinapis alba*) and radish (*Raphanus sativus*). They can be grown after a main (cereal) crop and trap nematodes, for example the beet cyst nematode and Columbian root knot nematode. When grown, nematodes hatch and are attracted to the roots. After entering the roots they cannot reproduce in the root due to a hypersensitive resistance reaction of the plant. Hence the nematode population is greatly reduced, by 70-99%, depending on species and cultivation time.

Other cover crops are used to attract natural predators of pests by providing elements of their habitat. This is a form of biological control known as habitat augmentation, but achieved with the use of cover crops. Findings on the relationship between cover crop presence and predator/pest population dynamics have been mixed, pointing toward the need for detailed information on specific cover crop types and management practices to best complement a given integrated pest management strategy. For example, the predator mite *Euseius tularensis* (Congdon) is known to help control the pest citrus thrips in Central California citrus orchards. Researchers found that the planting of several different leguminous cover crops (such as bell bean, woollypod vetch, New Zealand white clover, and Austrian winter pea) provided sufficient pollen as a feeding source to cause a seasonal increase in *E. tularensis* populations, which with good timing could potentially introduce enough predatory pressure to reduce pest populations of citrus thrips.

Diversity and Wildlife

Although cover crops are normally used to serve one of the above discussed purposes, they often simultaneously improve farm habitat for wildlife. The use of cover crops adds at least one more dimension of plant diversity to a cash crop rotation. Since the cover crop is typically not a crop of value, its management is usually less intensive, providing a window of "soft" human influence on the farm. This relatively "hands-off" management, combined with the increased on-farm heterogeneity created by the establishment of cover crops, increases the likelihood that a more complex trophic structure will develop to support a higher level of wildlife diversity.

In one study, researchers compared arthropod and songbird species composition and field use between conventionally and cover cropped cotton fields in the Southern United States. The cover cropped cotton fields were planted to clover, which was left to grow in between cotton rows throughout the early cotton growing season (stripcover cropping). During the migration and breeding season, they found that songbird densities were 7–20 times higher in the cotton fields with integrated clover cover crop than in the conventional cotton fields. Arthropod abundance and biomass was also higher in the clover cover cropped fields throughout much of the songbird breeding season, which was attributed to an increased supply of flower nectar from the clover. The clover cover crop enhanced songbird habitat by providing cover and nesting sites, and an increased food source from higher arthropod populations.

TILLAGE

Cultivating after an early rain.

Tillage is the agricultural preparation of soil by mechanical agitation of various types, such as digging, stirring, and overturning. Examples of human-powered tilling methods using hand tools include shovelling, picking, mattock work, hoeing, and raking. Examples of draft-animal-powered or mechanized work include ploughing (overturning with moldboards or chiseling with chisel shanks), rototilling, rolling with cultipackers or other rollers, harrowing, and cultivating with cultivator shanks (teeth). Small-scale gardening and farming, for household food production or small business production, tends to use the smaller-scale methods, whereas medium- to large-scale farming tends to use the larger-scale methods.

Tillage that is deeper and more thorough is classified as primary, and tillage that is shallower and sometimes more selective of location is secondary. Primary tillage such as ploughing tends to produce a rough surface finish, whereas secondary tillage tends to produce a smoother surface finish, such as that required to make a good seedbed for many crops. Harrowing and rototilling often combine primary and secondary tillage into one operation.

"Tillage" can also mean the land that is tilled. The word "cultivation" has several senses that overlap substantially with those of "tillage". In a general context, both can refer to agriculture. Within agriculture, both can refer to any kind of soil agitation. Additionally, "cultivation" or "cultivating" may refer to an even narrower sense of shallow, selective secondary tillage of row crop fields that kills weeds while sparing the crop plants.

Tilling with Hungarian Grey cattles.

Tilling was first performed via human labor, sometimes involving slaves. Hoofed animals could also be used to till soil by trampling. The wooden plow was then invented. It could be pulled with human labor, or by mule, ox, elephant, water buffalo, or similar sturdy animal. Horses are generally unsuitable, though breeds such as the Clydesdale were bred as draft animals. The steel plow allowed farming in the American Midwest, where tough prairie grasses and rocks caused trouble. Soon after 1900, the farm tractor was introduced, which eventually made modern large-scale agriculture possible.

Primary and Secondary Tillage

Primary tillage is usually conducted after the last harvest, when the soil is wet enough to allow plowing but also allows good traction. Some soil types can be plowed dry. The objective of primary tillage is to attain a reasonable depth of soft soil, incorporate crop residues, kill weeds, and to aerate the soil. Secondary tillage is any subsequent tillage, in order to incorporate fertilizers, reduce the soil to a finer tilth, level the surface, or control weeds.

Reduced Tillage

Reduced tillage leaves between 15 and 30% crop residue cover on the soil or 500 to 1000 pounds per acre (560 to 1100 kg/ha) of small grain residue during the critical erosion period. This may involve the use of a chisel plow, field cultivators, or other implements.

Plough tilling the field.

Intensive Tillage

Intensive tillage leaves less than 15% crop residue cover or less than 500 pounds per acre (560 kg/ha) of small grain residue. This type of tillage is often referred to as conventional tillage, but as conservational tillage is now more widely used than intensive tillage (in the United States), it is often not appropriate to refer to this type of tillage as conventional. Intensive tillage often involves multiple operations with implements such as a mold board, disk, and/or chisel plow. After this, a finisher with a harrow, rolling basket, and cutter can be used to prepare the seed bed. There are many variations.

Conservation Tillage

Conservation tillage leaves at least 30% of crop residue on the soil surface, or at least 1,000 lb/ac (1,100 kg/ha) of small grain residue on the surface during the critical soil erosion period. This slows water movement, which reduces the amount of soil erosion. Additionally, conservation tillage has been found to benefit predatory arthropods that can enhance pest control. Conservation tillage also benefits farmers by reducing fuel consumption and soil compaction. By reducing the number of times the farmer travels over the field, farmers realize significant savings in fuel and labor. In most years since 1997, conservation tillage was used in US cropland more than intensive or reduced tillage.

However, conservation tillage delays warming of the soil due to the reduction of dark earth exposure to the warmth of the spring sun, thus delaying the planting of the next year's spring crop of corn.

- No-till: Never use a plow, disk, etc. ever again. Aims for 100% ground cover.

- Strip-Till: Narrow strips are tilled where seeds will be planted, leaving the soil in between the rows untilled.

- Mulch-Till.

- Rotational Tillage: Tilling the soil every two years or less often (every other year, or every third year, etc.).

- Ridge-Till.

Zone Tillage

Zone tillage is a form of modified deep tillage in which only narrow strips are tilled, leaving soil in between the rows untilled. This type of tillage agitates the soil to help reduce soil compaction problems and to improve internal soil drainage. It is designed to only disrupt the soil in a narrow strip directly below the crop row. In comparison to no-till, which relies on the previous year's plant residue to protect the soil and aides in postponement of the warming of the soil and crop growth in Northern climates, zone tillage creates approximately a strip approximately five inches wide that simultaneously breaks up plow pans, assists in warming the soil and helps to prepare a seedbed. When combined with cover crops, zone tillage helps replace lost organic matter, slows the deterioration of the soil, improves soil drainage, increases soil water and nutrient holding capacity, and allows necessary soil organisms to survive.

It has been successfully used on farms in the Midwest and West for over 40 years, and is currently used on more than 36% of the U.S. farmland.Some specific states where zone tillage is currently in practice are Pennsylvania, Connecticut, Minnesota, Indiana, Wisconsin, and Illinois.

Unfortunately, its use in the Northern Cornbelt states lacks consistent yield results; however, there is still interest in deep tillage within the agricultureindustry. In areas that are not well-drained, deep tillage may be used as an alternative to installing more expensive tile drainage.

Effects of Tillage

Positive

Plowing:

- Loosens and aerates the top layer of soil or horizon A, which facilitates planting the crop.
- Helps mix harvest residue, organic matter (humus), and nutrients evenly into the soil.
- Mechanically destroys weeds.
- Dries the soil before seeding (in wetter climates tillage aids in keeping the soil drier).
- When done in autumn, helps exposed soil crumble over winter through frosting and defrosting, which helps prepare a smooth surface for spring planting.

Negative

- Dries the soil before seeding.
- Soil loses nutrients, like nitrogen and fertilizer, and its ability to store water.
- Decreases the water infiltration rate of soil. (Results in more runoff and erosion since the soil absorbs water more slowly than before).
- Tilling the soil results in dislodging the cohesiveness of the soil particles thereby inducing erosion.
- Chemical runoff.

- Reduces organic matter in the soil.

- Reduces microbes, earthworms, ants, etc.

- Destroys soil aggregates.

- Compaction of the soil, also known as a tillage pan.

- Eutrophication (nutrient runoff into a body of water).

- Can attract slugs, cut worms, army worms, and harmful insects to the leftover residues.

- Crop diseases can be harbored in surface residues.

General Comments

- The type of implement makes the most difference, although other factors can have an effect.

- Tilling in absolute darkness (night tillage) might reduce the number of weeds that sprout following the tilling operation by half. Light is necessary to break the dormancy of some weed species' seed, so if fewer seeds are exposed to light during the tilling process, fewer will sprout. This may help reduce the amount of herbicides needed for weed control.

- Greater speeds, when using certain tillage implements (disks and chisel plows), lead to more intensive tillage (i.e., less residue is on the soil surface).

- Increasing the angle of disks causes residues to be buried more deeply. Increasing their concavity makes them more aggressive.

- Chisel plows can have spikes or sweeps. Spikes are more aggressive.

- Percentage residue is used to compare tillage systems because the amount of crop residue affects the soil loss due to erosion.

Alternatives to Tilling

Modern agricultural science has greatly reduced the use of tillage. Crops can be grown for several years without any tillage through the use of herbicides to control weeds, crop varieties that tolerate packed soil, and equipment that can plant seeds or fumigate the soil without really digging it up. This practice, called no-till farming, reduces costs and environmental change by reducing soil erosion and diesel fuel usage.

Site Preparation of Forest Land

Site preparation is any of various treatments applied to a site in order to ready it for seeding or planting. The purpose is to facilitate the regeneration of that site by the chosen method. Site preparation may be designed to achieve, singly or in any combination: Improved access, by reducing or rearranging slash, and amelioration of adverse forest floor, soil, vegetation, or other biotic factors. Site preparation is undertaken to ameliorate one or more constraints that would otherwise be likely to thwart the objectives of management. A valuable bibliography on the effects of soil temperature and site preparation on subalpine and boreal tree species has been prepared by McKinnon et al.

Site preparation is the work that is done before a forest area is regenerated. Some types of site preparation are burning.

Burning

Broadcast burning is commonly used to prepare clearcut sites for planting, e.g., in central British Columbia, and in the temperate region of North America generally.

Prescribed burning is carried out primarily for slash hazard reduction and to improve site conditions for regeneration; all or some of the following benefits may accrue:

- Reduction of logging slash, plant competition, and humus prior to direct seeding, planting, scarifying or in anticipation of natural seeding in partially cut stands or in connection with seed-tree systems.

- Reduction or elimination of unwanted forest cover prior to planting or seeding, or prior to preliminary scarification thereto.

- Reduction of humus on cold, moist sites to favour regeneration.

- Reduction or elimination of slash, grass, or brush fuels from strategic areas around forested land to reduce the chances of damage by wildfire.

Prescribed burning for preparing sites for direct seeding was tried on a few occasions in Ontario, but none of the burns was hot enough to produce a seedbed that was adequate without supplementary mechanical site preparation.

Changes in soil chemical properties associated with burning include significantly increased pH, which Macadam in the Sub-boreal Spruce Zone of central British Columbia found persisting more than a year after the burn. Average fuel consumption was 20 to 24 t/ha and the forest floor depth was reduced by 28% to 36%. The increases correlated well with the amounts of slash (both total and ≥7 cm diameter) consumed. The change in pH depends on the severity of the burn and the amount consumed; the increase can be as much as 2 units, a 100-fold change. Deficiencies of copper and iron in the foliage of white spruce on burned clear-cuts in central British Columbia might be attributable to elevated pH levels.

Even a broadcast slash fire in a clear-cut does not give a uniform burn over the whole area. Tarrant, for instance, found only 4% of a 140-ha slash burn had burned severely, 47% had burned lightly, and 49% was unburned. Burning after windrowing obviously accentuates the subsequent heterogeneity.

Marked increases in exchangeable calcium also correlated with the amount of slash at least 7 cm in diameter consumed. Phosphorus availability also increased, both in the forest floor and in the 0 cm to 15 cm mineral soil layer, and the increase was still evident, albeit somewhat diminished, 21 months after burning. However, in another study in the same Sub-boreal Spruce Zone found that although it increased immediately after the burn, phosphorus availability had dropped to below pre-burn levels within 9 months.

Nitrogen will be lost from the site by burning, though concentrations in remaining forest floor were found by Macadam to have increased in two out of six plots, the others showing decreases. Nutrient losses may be outweighed, at least in the short term, by improved soil microclimate through the reduced thickness of forest floor where low soil temperatures are a limiting factor.

The *Picea/Abies* forests of the Alberta foothills are often characterized by deep accumulations of organic matter on the soil surface and cold soil temperatures, both of which make reforestation difficult and result in a general deterioration in site productivity; Endean and Johnstonedescribe experiments to test prescribed burning as a means of seedbed preparation and site amelioration on representative clear-felled *Picea/Abies*areas. Results showed that, in general, prescribed burning did not reduce organic layers satisfactorily, nor did it increase soil temperature, on the sites tested. Increases in seedling establishment, survival, and growth on the burned sites were probably the result of slight reductions in the depth of the organic layer, minor increases in soil temperature, and marked improvements in the efficiency of the planting crews. Results also suggested that the process of site deterioration has not been reversed by the burning treatments applied.

Ameliorative Intervention

Slash weight (the oven-dry weight of the entire crown and that portion of the stem less than four inches in diameter) and size distribution are major factors influencing the forest fire hazard on harvested sites. Forest managers interested in the application of prescribed burning for hazard reduction and silviculture, were shown a method for quantifying the slash load by Kiil. In west-central Alberta, he felled, measured, and weighed 60 white spruce, graphed (a) Slash weight per merchantable unit volume against diameter at breast height (dbh), and (b) Weight of fine slash (<1.27 cm) also against dbh, and produced a table of slash weight and size distribution on one acre of a hypothetical stand of white spruce. When the diameter distribution of a stand is unknown, an estimate of slash weight and size distribution can be obtained from average stand diameter, number of trees per unit area, and merchantable cubic foot volume. The sample trees in Kiil's study had full symmetrical crowns. Densely growing trees with short and often irregular crowns would probably be overestimated; open-grown trees with long crowns would probably be underestimated.

The need to provide shade for young out plants of Engelmann spruce in the high Rocky Mountains is emphasized by the U.S. Forest Service. Acceptable planting spots are defined as microsites on the north and east sides of down logs, stumps, or slash, and lying in the shadow cast by such material. Where the objectives of management specify more uniform spacing, or higher densities, than obtainable from an existing distribution of shade-providing material, redistribution or importing of such material has been undertaken.

Access

Site preparation on some sites might be done simply to facilitate access by planters, or to improve access and increase the number or distribution of microsites suitable for planting or seeding.

Wang et al. determined field performance of white and black spruces 8 and 9 years after out planting on boreal mixed wood sites following site preparation (Donaren disc trenching versus no trenching) in 2 plantation types (open versus sheltered) in south-eastern Manitoba. Donaren trenching slightly reduced the mortality of black spruce but significantly increased the mortality of white spruce. Significant difference in height was found between open and sheltered plantations for black spruce but not for white spruce, and root collar diameter in sheltered plantations was significantly larger than in open plantations for black spruce but not for white spruce. Black spruce open plantation had significantly smaller volume (97 cm³) compared with black spruce sheltered (210 cm³), as well as white spruce open (175 cm³) and sheltered (229 cm³) plantations. White spruce open plantations also had smaller volume than white spruce sheltered plantations. For transplant stock, strip plantations had a significantly higher volume (329 cm³) than open plantations (204 cm³). Wang et al. recommended that sheltered plantation site preparation should be used.

Mechanical

Up to 1970, no "sophisticated" site preparation equipment had become operational in Ontario, but the need for more efficacious and versatile equipment was increasingly recognized. By this time, improvements were being made to equipment originally developed by field staff, and field testing of equipment from other sources was increasing.

According to J. Hall, in Ontario at least, the most widely used site preparation technique was post-harvest mechanical scarification by equipment front-mounted on a bulldozer (blade, rake, V-plow, or teeth), or dragged behind a tractor (Imsett or S.F.I. scarifier, or rolling chopper). Drag type units designed and constructed by Ontario's Department of Lands and Forests used anchor chain or tractor pads separately or in combination, or were finned steel drums or barrels of various sizes and used in sets alone or combined with tractor pad or anchor chain units.

J. Hall's report on the state of site preparation in Ontario noted that blades and rakes were found to be well suited to post-cut scarification in tolerant hardwood stands for natural regeneration of yellow birch. Plows were most effective for treating dense brush prior to planting, often in conjunction with a planting machine. Scarifying teeth, e.g., Young's teeth, were sometimes used to prepare sites for planting, but their most effective use was found to be preparing sites for seeding, particularly in backlog areas carrying light brush and dense herbaceous growth. Rolling choppers found application in treating heavy brush but could be used only on stone-free soils. Finned drums were commonly used on jack pine–spruce cutovers on fresh brushy sites with a deep duff layer and heavy slash, and they needed to be teamed with a tractor pad unit to secure good distribution of the slash. The S.F.I. scarifier, after strengthening, had been "quite successful" for 2 years, promising trials were under way with the cone scarifier and barrel ring scarifier, and development had begun on a new flail scarifier for use on sites with shallow, rocky soils. Recognition of the need to become more effective and efficient in site preparation led the Ontario Department of Lands and Forests to adopt the policy of seeking and obtaining for field testing new equipment from Scandinavia and elsewhere that seemed to hold promise for Ontario conditions, primarily in the north. Thus, testing was begun of the Brackekultivator from Sweden and the Vako-Visko rotary furrower from Finland.

Mounding

Site preparation treatments that create raised planting spots have commonly improved out plant performance on sites subject to low soil temperature and excess soil moisture. Mounding can certainly have a big influence on soil temperature. Draper et al., for instance, documented this as well as the effect it had on root growth of out plants.

The mounds warmed up quickest, and at soil depths of 0.5 cm and 10 cm averaged 10 and 7 °C higher, respectively, than in the control. On sunny days, daytime surface temperature maxima on the mound and organic mat reached 25 °C to 60 °C, depending on soil wetness and shading. Mounds reached mean soil temperatures of 10 °C at 10 cm depth 5 days after planting, but the control did not reach that temperature until 58 days after planting. During the first growing season, mounds had 3 times as many days with a mean soil temperature greater than 10 °C than did the control microsites.

Draper et al. mounds received 5 times the amount of photo synthetically active radiation (PAR) summed over all sampled microsites throughout the first growing season; the control treatment consistently received about 14% of daily background PAR, while mounds received over 70%. By November, fall frosts had reduced shading, eliminating the differential. Quite apart from its effect on temperature, incident radiation is also important photo synthetically. The average control microsite was exposed to levels of light above the compensation point for only 3 hours, i.e., one-quarter of the daily light period, whereas mounds received light above the compensation point for 11 hours, i.e., 86% of the same daily period. Assuming that incident light in the 100-600 $\mu Em^{-2}s^{-1}$ intensity range is the most important for photosynthesis, the mounds received over 4 times the total daily light energy that reached the control microsites.

Orientation of Linear Site Preparation

With linear site preparation, orientation is sometimes dictated by topography or other considerations, but the orientation can often be chosen. It can make a difference. A disk-trenching experiment in the Sub-boreal Spruce Zone in interior British Columbia investigated the effect on growth of young out plants (lodgepole pine) in 13 microsite planting positions: berm, hinge, and trench in each of north, south, east, and west aspects, as well as in untreated locations between the furrows. Tenth-year stem volumes of trees on south, east and west-facing microsites were significantly greater than those of trees on north-facing and untreated microsites. However, planting spot selection was seen to be more important overall than trench orientation.

In a Minnesota study, the N–S strips accumulated more snow but snow melted faster than on E–W strips in the first year after felling. Snow-melt was faster on strips near the centre of the strip-felled area than on border strips adjoining the intact stand. The strips, 50 feet (15.24 m) wide, alternating with uncut strips 16 feet (4.88 m) wide, were felled in a *Pinus resinosa* stand, aged 90 to 100 years.

GREEN MANURING

Green manuring is defined as the growing of green manure crops & then turning off these crops directly in the field by ploughing the field so as to make the field richer in nitrogen which is the most deficient nutrient of the soil. Green manuring crops help in improving the structure of soil & also increases its physical properties. One of the main objectives of the green manuring is to increase the content of nitrogen in the soil to increase the crop production. The green manuring can be practised as in-situ or green leaf manuring. Green in-situ manuring refers to the growing of green manuring crops in the border rows or as intercrops along with the main crops for example: Sunn hemp, Cowpea, Dhaincha, Berseem, Green gram etc. whereas green leaf manuring is the collection of green leaves from outside places such as waste or forestlands; for example collection of wild dhaincha leaves & then incorporating them into the crop field for improving the soil properties. Most of the green manuring crops are incorporated in the fields after 6 to 8 weeks of sowing with the application of water so that they should be easily incorporated into the soil. The flowering stage of the green manuring crops is the best time for incorporation of these crops.

Different Green Manuring Crops

The different green manuring crops are listed in detail below:

- Crotolaria juncea (or Sunnhemp): It is a green manuring crop with vigorous cropping habits & it does not withstand the waterlogging. Its seed rate is 25 -30 kg. per hectare. It is incorporated into the soil after a period of 9 or 10 weeks of sowing. The quantity of nitrogen fixed by this crop is very good i.e.80 to 130 kg/ha. It contains about 2.80 to 3.15% nitrogen content percent.

- Sesbania aculeate (Daincha): It is a quick growing succulent crop which can be grown in any conditions of soil & climate. It can be incorporated into the soil after 8 to 10 weeks after sowing & its recommended seed rate is 20 to 25 kg/ha. The quantity of nitrogen fixed by this crop is 130 to 18 5 kg/ha & N content in it is 2.55 to 3.21%.

- Sesbania rostrata: It is different from S. Acumeate in the sense that it has nodules on both stem & root. It grows well under waterlogged conditions. It's recommended seed rate is 35 to 40 kg/ha & the seed should be sown after treating it with sulphuric acid for 15 minutes for effective germination. It can fix a nitrogen dose of 170 to 220 kg/ha within a period of 8 to 10 weeks in soil & contains 3.20 to 3.37% N content.

- Sesbania species: It is another species of daincha which is also a good green manuring crop. It fixes 115 to 160 kg nitrogen/ha in the soil & contains 2.29 to 3.1% N content in it.

- Tephrpsia purpurea (Wild indigo): It is a slow growing green manure crop which is not eaten by cattle & it does not withstand waterlogging conditions for a long time. Due to these reasons, it is most suitable for light soils green manuring. It's effective seed rate is 20 to 25 kg/ha & it fixes nitrogen @ 70 to 115 kg/ha. It contains 2.90 to 3.22 % N content.

- Glyricidia maculata: It is a shrub which grows tall & used as green leaf manure in the field. It can be kept low by topping or pruning at a height of 2 to 3 meters from the ground as it has a tall tree like growing habits. It requires 2 to 3 annual pruning & can give 5 to 10 kg leaves/plant annually after 2 years of growth. It contains 2.25 to 2.76 % N content.

- Cassia pistula: It is a weed which is used for green manuring. It contains 0.75 to 1.22 % nitrogen content in it. Its stems & leaves are used for green manuring when the plant is in the flowering stage.

- Cowpea or Vigna sinensis: It fixes about 200 kg nitrogen/ha in soil due to symbiotic nitrogen fixation by Rhizobium.

- Green gram or Vigna radiata: It fixes about 220 kg nitrogen/ha in the soil which is very high in regards that you can also obtain a good production of a green gram as well.

- Berseem or Trifoilium alexandrinum: The fodder crop of berseem fixes about 280 to 350 kg nitrogen/ha.

- Some of the other legume green manuring crops are Soybean, Alfa-Alfa, Clusterbean etc. which also fixes effective nitrogen from the atmosphere into the soil by the process of symbiotic nitrogen fixation by a bacterium named Rhizobium.

Advantages of Green Manuring

There are many advantages of green manuring which are defined below:

- It will lead in building the soil structure & improving the soil physical properties like soil aeration, water holding capacity of the soil, soil bulk density etc.

- This will also help in bringing the nutrients into the upper layers so that they should become available for the plant absorption from the lower layers.

- Green manuring also fulfils the crops need of nitrogen which is the most deficient nutrient in the soil these days.

- It will also increase the soil availability of different micro and macronutrient like those of calcium, phosphorus, potash, magnesium etc.

Types of Green Manure

Common crops used for green manure include soybeans, clover, and rye, but a wide variety of plants can be used. Each type of crop provides certain benefits. Most plants improve the nitrogen levels in your soil once they are tilled in.

A green manure crop of alfalfa can send roots down as far as 60 feet, which pulls nutrients from the depths up to the surface to benefit your next season's crop. Winter tare and hairy vetch have extensive root systems that prevent erosion and long leafy vines that grow in the winter, protecting the surface of your soil—serving as a built-in mulch that won't blow away during winter storms.

Woody pod vetch is super hardy and robust and helps eliminate weeds by crowding them out. Some crops, such as lupine, have relatively quick life cycles, so you can perform green manure treatments in as short as eight-week cycles. Agricultural mustard matures in as little as six weeks.

If you can grow them in your area, legumes like peas and fava beans are especially good for adding nitrogen to soil, since the nutrient is contained in the plants' roots.

You also can make a simple green manure by collecting green leaves and twigs from plants grown in wastelands, forests, and other areas, although growing green manure crops is more common in organic farming.

Growing the cover crops yourself also ensures that you'll be getting organic material; it's hard to know whether wild plants in some areas have been affected by disease or even pesticide sprayings or applications.

Tilling Green Manure

Typically, you till your green manure crop into the soil just before it flowers, which prevents the plants from going to seed and growing again after you dig them into the soil.

Tilling before plants flower is critical; most of a plant's nitrogen is gone once it's reached the flowering stage. With some plants, you may want to let a few go to flower so that you can better attract beneficial pollinating insects.

For larger plants, you may need to break them down first with hand tools or a weed whacker. Allow two to three weeks after tilling so that the organic material can decompose, and then plant your desired crop.

Some cover crops are easier to grow and till than others. If you don't have industrial farming equipment, most vetches, tares, and clovers are difficult to till.

Be patient when figuring out what crops work best with your soil. Some crops do better in hard soil, others are more suited for light or dry soils, and your soil might even differ from place to place on your property. A few seasons may pass before you get just the right balance of cover crops, tilling, and replanting.

References

- Soil-management, soils-portal: fao.org, Retrieved 15 February, 2019

- Eka Putri, E.; Kameswara Rao, N.S.V.; Mannan, M.A. (2012). "Evaluation of Modulus of Elasticity and Modulus of Subgrade Reaction of Soils Using CBR Test". Journal of Civil Engineering Research. 2 (1): 34–40. Doi:10.5923/j.jce.20120201.05

- Introduction-to-soils-managing-soils: extension.psu.edu, Retrieved 16 March, 2019

- Kick, Chris (18 February 2014). "New soil test measures soil health - Farm and Dairy". Farmanddairy.com. Archived from the original on 1 December 2017. Retrieved 26 April 2018.

- Soil-survey-objectives-and-types, soil: biologydiscussion.com, Retrieved 17 April, 2019

- Mitchell, J.K., and Soga, K. (2005) Fundamentals of soil behavior, Third edition, John Wiley and Sons, Inc., ISBN 978-0-471-46302-3

- How-to-condition-soil, soil-fertilizers, garden-how-to: gardeningknowhow.com, Retrieved 18 May, 2019

- "Healthy Soil". www.highbrixgardens.com. Archived from the original on 19 December 2016. Retrieved 26 April 2018

- Green-manuring: abcofagri.com, Retrieved 19 June, 2019

- Mäder, Paul; et al. (2000). "Arbuscular mycorrhizae in a long-term field trial comparing low-input (organic, biological) and high-input (conventional) farming systems in a crop rotation". Biology and Fertility of Soils. 31 (2): 150–156. Doi:10.1007/s003740050638

- What-is-green-manure-2538194: thebalancesmb.com, Retrieved 20 July, 2019

Soil Pollution and Conservation

The presence of high concentrations of toxic chemicals in soil that can pose risk to human health and the ecosystem is known as soil pollution. The prevention of soil loss from erosion and reduced soil fertility caused by acidification is known as soil conservation. The topics elaborated in this chapter will help in gaining a better perspective about soil pollution and soil conservation.

SOIL POLLUTION

Soil pollution refers to anything that causes contamination of soil and degrades the soil quality. It occurs when the pollutants causing the pollution reduce the quality of the soil and convert the soil inhabitable for microorganisms and macro organisms living in the soil.

Soil contamination or soil pollution can occur either because of human activities or because of natural processes. However, mostly it is due to human activities. The soil contamination can occur due to the presence of chemicals such as pesticides, herbicides, ammonia, petroleum hydrocarbons, lead, nitrate, mercury, naphthalene, etc in an excess amount.

The primary cause of soil pollution is a lack of awareness in general people. Thus, due to many different human activities such as overuse of pesticides the soil will lose its fertility. Moreover, the presence of excess chemicals will increase the alkalinity or acidity of soil thus degrading the soil quality. This will in turn cause soil erosion. This soil erosion refers to soil pollution.

Causes of Soil Pollution

Soil pollution can be natural or due to human activity. However, it mostly boils down to the activities of the human that causes the majority of soil pollution such as heavy industries, or pesticides in agriculture.

Pesticides

Before World War II, the chemical nicotine chemical present in the tobacco plants was used as the pest controlling substance in agricultural practices. However, DDT was found to be extremely useful for malaria control and as pest control of many insects during World War II. Therefore, it was used for controlling many diseases.

Hence, post-war, people started using it as pest control in agriculture for killing rodents, weeds, insects, etc and avoiding the damages due to these pests. However, everyone gradually the adverse effects of this chemical which led to the ban of this chemical in many parts of the world including India.

Moreover, pests became resistance to DDT due to the chemicals regular use. Hence this led to the introduction of other harmful chemicals such as Aldrin and Dieldrin. Pesticides are synthetic toxic chemicals that definitely kill different types of pests and insects causing damage to agriculture but it has many ecological repercussions.

They are generally insoluble in water and non-biodegradable. Therefore, these chemicals will not gradually decompose and keep on accumulating in the soil. Therefore, the concentration of these chemicals will increase when the transfer of these chemicals take place from lower to higher trophic level via the food chain. Hence, it will cause many metabolic and physiological disorders in humans.

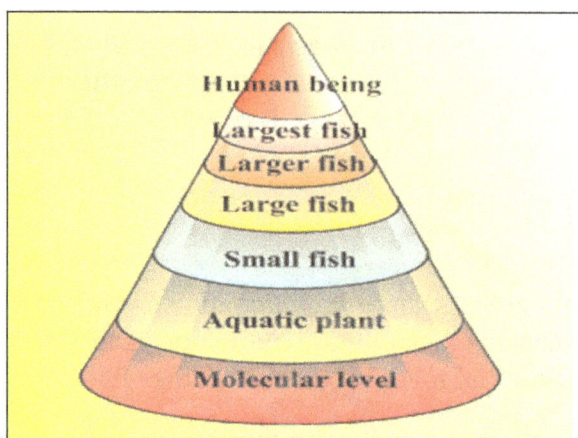

Trophic level concentration of pollutants increases 10 fold at each trophic level.

Chlorinated Organic Toxins

The harmful effect of DDT and other chemicals led to the introduction of less persistent organic and more-biodegradable substance such as carbamates and organophosphates. However, these chemicals act as harmful toxins for nerves, hence they are more dangerous to humans. It led to pesticides related to the death of field workers in some agricultural fields.

Herbicides

Slowly, the industries began production of herbicides like sodium arsenite (Na_3AsO_3), sodium chlorate ($NaClO_3$), etc. Herbicides can decompose in a span of few months. However, even they affect the environment and are not environmental friendly. Even though they are not as harmful as organo-chlorides but most of the herbicides are toxic. They are known to cause birth defects.

Furthermore, research suggests that spraying herbicides causes more insect attack and diseases of plants in comparison to manual weeding. One thing to note here is all the above factors occupy just a small portion of the causes. Majority of the causes is related to manufacturing activities in chemical and industrial processes that are released in nature or environment.

Inorganic Fertilizers

Excessive use of inorganic nitrogen fertilizers leads to acidification of soil and contaminate the agricultural soil. Also known as agrochemical pollution.

Industrial Pollution

The incorrect way of chemical waste disposal from different types of industries can cause contamination of soil. Human activities like this have led to acidification of soil and contamination due to the disposal of industrial waste, heavy metals, toxic chemicals, dumping oil and fuel, etc.

Inferior Irrigation Practices

Poor irrigation methods increase the soil salinity. Moreover, excess watering, improper maintenance of canals and irrigation channels, lack of crop rotation and intensive farming gradually decreases the quality of soil over time and cause degradation of land.

Solid Waste

Disposal of plastics, cans, and other solid waste falls into the category of soil pollution. Disposal of electrical goods such as batteries causes an adverse effect on the soil due to the presence of harmful chemicals. For instance, lithium present in batteries can cause leaching of soil.

Urban Activities

Lack of proper waste disposal, regular constructions can cause excessive damage to the soil due to lack of proper drainage and surface run-off. These waste disposed of by humans contain chemical waste from residential areas. Moreover leaking of sewerage system can also affect soil quality and cause soil pollution by changing the chemical composition of the soil.

After-effects of Soil Pollution

Soil pollution is not only the problem in India but it is a global problem. It causes harmful effect on the soil and the environment at large. Contamination of soil will decrease the agricultural output of a land. Major soil pollution after effects are given below.

Inferior Crop Quality

It can decrease the quality of the crop. Regular use of chemical fertilizers, inorganic fertilizers, pesticides will decrease the fertility of the soil at a rapid rate and alter the structure of the soil. This will lead to decrease in soil quality and poor quality of crops. Over the time the soil will become less productive due to the accumulation of toxic chemicals in large quantity.

Harmful Effect on Human Health

It will increase the exposure to toxic and harmful chemicals thus increasing health threats to people living nearby and on the degraded land. Living, working or playing in the contaminated soil can lead to respiratory diseases, skin diseases, and other diseases. Moreover, it can cause other health problems.

Water Sources Contamination

The surface run-off after raining will carry the polluted soil and enter into different water resource.

Thus, it can cause underground water contamination thereby causing water pollution. This water after contamination is not fit for human as well as animal use due to the presence of toxic chemicals.

Negative Impact on Ecosystem and Biodiversity

Soil pollution can cause an imbalance of the ecosystem of the soil. The soil is an important habitat and is the house of different type of microorganisms, animals, reptiles, mammals, birds, and insects. Thus, soil pollution can negatively impact the lives of the living organisms and can result in the gradual death of many organisms. It can cause health threats to animals grazing in the contaminated soil or microorganisms residing in the soil.

Therefore, human activities are responsible for the majority of the soil pollution. We as humans buy things that are harmful and not necessary, use agricultural chemicals (fertilizers, pesticides, herbicides, etc.), drop waste here and there. Without being aware we harm our own environment.

Therefore, it is very important to educate people around you the importance of environment if they are not aware. Prevention of soil erosion will help to cease soil pollution. Thus, it is our small steps and activities that can help us to achieve a healthier planet for us. Therefore, it is essential for industries, individuals and businesses to understand the importance of soil and prevent soil pollution and stop the devastation caused to plant and animal life.

SOIL RETROGRESSION AND DEGRADATION

Soil retrogression and degradation are two regressive evolution processes associated with the loss of equilibrium of a stable soil. Retrogression is primarily due to soil erosion and corresponds to a phenomenon where succession reverts the land to its natural physical state. Degradation is an evolution, different from natural evolution, related to the local climate and vegetation. It is due to the replacement of primary plant communities (known as climax vegetation) by the secondary communities. This replacement modifies the humuscomposition and amount, and affects the formation of the soil. It is directly related to human activity. Soil degradation may also be viewed as any change or ecological disturbance to the soil perceived to be deleterious or undesirable.

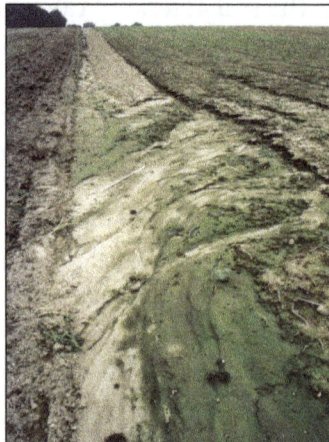

Intensive tillage result on soil degradation.

At the beginning of soil formation, the bare rock out crops is gradually colonized by pioneer species (lichens and mosses). They are succeeded by herbaceous vegetation, shrubs and finally forest. In parallel, the first humus-bearing horizon is formed (the A horizon), followed by some mineral horizons (B horizons). Each successive stage is characterized by a certain association of soil/vegetation and environment, which defines an ecosystem.

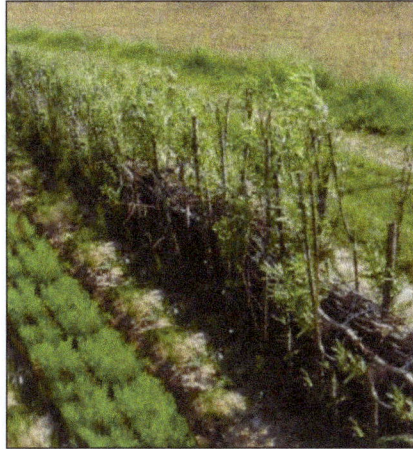

Willow hedge strengthened with fascines for the limitation of runoff, northern France.

After a certain time of parallel evolution between the ground and the vegetation, a state of steady balance is reached. This stage of development is called climax by some ecologists and "natural potential" by others. Succession is the evolution towards climax. Regardless of its name, the equilibrium stage of primary succession is the highest natural form of development that the environmental factors are capable of producing.

The cycles of evolution of soils have very variable durations, between tens, hundreds, or thousands of years for quickly evolving soils (A horizon only) to more than a million years for slowly developing soils. The same soil may achieve several successive steady state conditions during its existence, as exhibited by the Pygmy forest sequence in Mendocino County, California. Soils naturally reach a state of high productivity, from which they naturally degrade as mineral nutrients are removed from the soil system. Thus older soils are more vulnerable to the effects of induced retrogression and degradation.

Ecological Factors Influencing Soil Formation

There are two types of ecological factors influencing the evolution of a soil (through alteration and humification). These two factors are extremely significant to explain the evolution of soils of short development.

- A first type of factor is the average climate of an area and the vegetation which is associated (biome).

- A second type of factor is more local, and is related to the original rock and local drainage. This type of factor explains appearance of specialized associations (ex peat bogs).

Biorhexistasy Theory

The destruction of the vegetation implies the destruction of evoluted soils, or a regressive evolution.

Cycles of succession-regression of soils follow one another within short intervals of time (human actions) or long intervals of time (climate variations).

The climate role in the deterioration of the rocks and the formation of soils lead to the formulation of the theory of the biorhexistasy.

- In wet climate, the conditions are favorable to the deterioration of the rocks (mostly chemically), the development of the vegetation and the formation of soils; this period favorable to life is called biostasy.

- In dry climate, the rocks exposed are mostly subjected to mechanical disintegration which produces coarse detrital materials: this is referred to as rhexistasy.

Perturbations of the Balance of a Soil

When the state of balance, characterized by the ecosystem climax is reached, it tends to be maintained stable in the course of time. The vegetation installed on the ground provides the humus and ensures the ascending circulation of the matters. It protects the ground from erosion by playing the role of barrier (for example, protection from water and wind). Plants can also reduce erosion by binding the particles of the ground to their roots.

A disturbance of climax will cause retrogression, but often, secondary succession will start to guide the evolution of the system after that disturbance. Secondary succession is much faster than primary because the soil is already formed, although deteriorated and needing restoration as well.

However, when a significant destruction of the vegetation takes place (of natural origin such as an avalanche or human origin), the disturbance undergone by the ecosystem is too important. In this latter case, erosion is responsible for the destruction of the upper horizons of the ground, and is at the origin of a phenomenon of reversion to pioneer conditions. The phenomenon is called retrogression and can be partial or total (in this case, nothing remains beside bare rock). For example, the clearing of an inclined ground, subjected to violent rains, can lead to the complete destruction of the soil. Man can deeply modify the evolution of the soils by direct and brutal action, such as clearing, abusive cuts, forest pasture, litters raking. The climax vegetation is gradually replaced and the soil modified (example: replacement of leafy tree forests by moors or pines plantations). Retrogression is often related to very old human practices.

Influence of Human Activity

Soil erosion is the main factor for soil degradation and is due to several mechanisms: water erosion, wind erosion, chemical degradation and physical degradation.

Erosion is strongly related to human activity. For example, roads which increase impermeable surfaces lead to streaming and ground loss. Agriculture also accelerates soil erosion (increase of field size, correlated to hedges and ditches removal). Meadows are in regression to the profit of plowed lands. Spring cultures (sunflower, corn, beet) surfaces are increasing and leave the ground naked in winter. Sloping grounds are gradually colonized by vine. Lastly, use of herbicides leaves the ground naked between each crop. New cultural practices, such as mechanization also increases the risks of erosion. Fertilization by mineral manures rather than organic manure gradually destructure the soil. Many scientists observed a gradual decrease of soil organic matter content in soils, as well as a decrease of soil biological activity (in particular, in relation to chemical uses). Lastly, deforestation, in particular, is responsible for degradation of forest soils.

Agriculture increases the risk of erosion through its disturbance of vegetation by way of:

- Overgrazing of animals,
- Monoculture planting,
- Row cropping,
- Tilling or plowing,
- Crop removal,
- Land-use conversion.

Consequences of Soil Regression and Degradation

- Yields impact: Recent increases in the human population have placed a great strain on the world's soil systems. More than 6 billion people are now using about 38% of the land area of the Earth to raise crops and livestock. Many soils suffer from various types of degradation, that can ultimately reduce their ability to produce food resources. Slight degradation refers to land where yield potential has been reduced by 10%, moderate degradation refers to a yield decrease from 10-50%. Severely degraded soils have lost more than 50% of their potential. Most severely degraded soils are located in developing countries.

- Natural disasters: Natural disasters such as mud flows, floods are responsible for the death of many living beings each year.

- Deterioration of the water quality: The increase in the turbidity of water and the contribution of nitrogen and of phosphorus can result in eutrophication. Soils particles in surface waters are also accompanied by agricultural inputs and by some pollutants of industrial, urban and road origin (such as heavy metals). The ecological impact of agricultural inputs (such as weed killer) is known but difficult to evaluate because of the multiplicity of the products and their broad spectrum of action.

- Biological diversity: Soil degradation may involve perturbation of microbial communities, disappearance of the climax vegetation and decrease in animal habitat, thus leading to a biodiversity loss and animal extinction.

- Economic loss: The estimated costs for land degradation are US $40 billion per year. This figure does not take into account negative externalities such as the increased use of fertilizers, loss of biodiversity and loss of unique landscapes.

Soil Enhancement, Rebuilding and Regeneration

Problems of soil erosion can be fought, and certain practices can lead to soil enhancement and rebuilding. Even though simple, methods for reducing erosion are often not chosen because these practices outweigh the short-term benefits. Rebuilding is especially possible through the improvement of soil structure, addition of organic matter and limitation of runoff. However, these techniques will never totally succeed to restore a soil (and the fauna and flora associated to it) that took more than 1000 years to build up. Soil regeneration is the reformation of degraded soil through biological, chemical, and or physical processes.

When productivity declined in the low-clay soils of northern Thailand, farmers initially responded by adding organic matter from termite mounds, but this was unsustainable in the long-term. Scientists experimented with adding bentonite, one of the smectite family of clays, to the soil. In field trials, conducted by scientists from the International Water Management Institute in cooperation with Khon Kaen Universityand local farmers, this had the effect of helping retain water and nutrients. Supplementing the farmer's usual practice with a single application of 200 kg bentonite per rai (6.26 rai = 1 hectare) resulted in an average yield increase of 73%. More work showed that applying bentonite to degraded sandy soils reduced the risk of crop failure during drought years.

In 2008, three years after the initial trials, IWMI scientists conducted a survey among 250 farmers in northeast Thailand, half who had applying bentonite to their fields and half who had not. The average output for those using the clay addition was 18% higher than for non-clay users. Using the clay had enabled some farmers to switch to growing vegetables, which need more fertile soil. This helped to increase their income. The researchers estimated that 200 farmers in northeast Thailand and 400 in Cambodia had adopted the use of clays, and that a further 20,000 farmers were introduced to the new technique.

SOIL EROSION

Soil erosion is a process that involves the wearing away of the topsoil. The process involves the loosening of the soil particles, blowing or washing away of the soil particles, and either ends up in the valley and faraway lands or washed away to the oceans by rivers and streams. Soil erosion is a natural process which has increasingly been exacerbated by human activities such as agriculture and deforestation.

The wearing away of the topsoil is driven by erosion agents including the natural physical forces of wind and water, each contributing a substantial quantity of soil loss annually. Farming activities such as tillage also significantly contribute to soil erosion.

Dry-land-muddy-volcanes-dry-land-soil-erosion.

Thus, soil erosion is a continuous process and may occur either at a relatively unnoticed rate or an alarming rate contributing to copious loss of the topsoil. The outcomes of soil erosion are reduced agricultural productivity, ecological collapse, soil degradation, and the possibility of desertification.

Causes of Soil Erosion

All soils undergo soil erosion, but some are more vulnerable than others due to human activities and other natural causal factors. The severity of soil erosion is also dependent on the soil type and the presence of vegetation cover. Here are few of the major causes of soil erosion.

Water Erosion

The widespread occurrence of water erosion combined with the severity of on-site and off-site impacts have made water erosion the focus of soil conservation efforts in Ontario.

The rate and magnitude of soil erosion by water is controlled by the following factors given below.

Rainfall and Runoff

The greater the intensity and duration of a rainstorm, the higher the erosion potential. The impact of raindrops on the soil surface can break down soil aggregates and disperse the aggregate material. Lighter aggregate materials such as very fine sand, silt, clay and organic matter are easily removed by the raindrop splash and runoff water; greater raindrop energy or runoff amounts are required to move larger sand and gravel particles.

Soil movement by rainfall (raindrop splash) is usually greatest and most noticeable during short-duration, high-intensity thunderstorms. Although the erosion caused by long-lasting and less-intense storms is not usually as spectacular or noticeable as that produced during thunderstorms, the amount of soil loss can be significant, especially when compounded over time.

The erosive force of wind on an open field.

Surface water runoff occurs whenever there is excess water on a slope that cannot be absorbed into the soil or is trapped on the surface. Reduced infiltration due to soil compaction, crusting or freezing increases the runoff. Runoff from agricultural land is greatest during spring months when the soils are typically saturated, snow is melting and vegetative cover is minimal.

Soil Erodibility

Soil erodibility is an estimate of the ability of soils to resist erosion, based on the physical characteristics of each soil. Texture is the principal characteristic affecting erodibility, but structure, organic matter and permeability also contribute. Generally, soils with faster infiltration rates, higher levels of organic matter and improved soil structure have a greater resistance to erosion. Sand, sandy loam and loam-textured soils tend to be less erodible than silt, very fine sand and certain clay-textured soils.

Tillage and cropping practices that reduce soil organic matter levels, cause poor soil structure, or result in soil compaction, contribute to increases in soil erodibility. As an example, compacted subsurface soil layers can decrease infiltration and increase runoff. The formation of a soil crust, which tends to "seal" the surface, also decreases infiltration. On some sites, a soil crust might decrease the amount of soil loss from raindrop impact and splash; however, a corresponding increase in the amount of runoff water can contribute to more serious erosion problems.

Past erosion also has an effect on a soil's erodibility. Many exposed subsurface soils on eroded sites tend to be more erodible than the original soils were because of their poorer structure and lower organic matter. The lower nutrient levels often associated with subsoils contribute to lower crop yields and generally poorer crop cover, which in turn provides less crop protection for the soil.

Slope Gradient and Length

The steeper and longer the slope of a field, the higher the risk for erosion. Soil erosion by water increases as the slope length increases due to the greater accumulation of runoff. Consolidation of small fields into larger ones often results in longer slope lengths with increased erosion potential, due to increased velocity of water, which permits a greater degree of scouring (carrying capacity for sediment).

Cropping and Vegetation

The potential for soil erosion increases if the soil has no or very little vegetative cover of plants and/or crop residues. Plant and residue cover protects the soil from raindrop impact and splash, tends to slow down the movement of runoff water and allows excess surface water to infiltrate.

The erosion-reducing effectiveness of plant and/or crop residues depends on the type, extent and quantity of cover. Vegetation and residue combinations that completely cover the soil and intercept all falling raindrops at and close to the surface are the most efficient in controlling soil erosion (e.g., forests, permanent grasses). Partially incorporated residues and residual roots are also important as these provide channels that allow surface water to move into the soil.

The effectiveness of any protective cover also depends on how much protection is available at various periods during the year, relative to the amount of erosive rainfall that falls during these periods. Crops that provide a full protective cover for a major portion of the year (e.g., alfalfa or winter cover crops) can reduce erosion much more than can crops that leave the soil bare for a longer period of time (e.g., row crops), particularly during periods of highly erosive rainfall such as spring and summer. Crop management systems that favour contour farming and strip-cropping techniques can further reduce the amount of erosion. To reduce most of the erosion on annual row-crop land, leave a residue cover greater than 30% after harvest and over the winter months, or inter-seed a cover crop (e.g., red clover in wheat, oats after silage corn).

Tillage Practices

The potential for soil erosion by water is affected by tillage operations, depending on the depth, direction and timing of plowing, the type of tillage equipment and the number of passes. Generally, the less the disturbance of vegetation or residue cover at or near the surface, the more effective the tillage practice in reducing water erosion. Minimum till or no-till practices are effective in reducing soil erosion by water.

Tillage and other practices performed up and down field slopes creates pathways for surface water runoff and can accelerate the soil erosion process. Cross-slope cultivation and contour farming techniques discourage the concentration of surface water runoff and limit soil movement.

Forms of Water Erosion

Sheet Erosion

Sheet erosion is the movement of soil from raindrop splash and runoff water. It typically occurs evenly over a uniform slope and goes unnoticed until most of the productive topsoil has been lost. Deposition of the eroded soil occurs at the bottom of the slope or in low areas. Lighter-coloured soils on knolls, changes in soil horizon thickness and low crop yields on shoulder slopes and knolls are other indicators.

The accumulation of soil and crop debris at the lower end of this field is an indicator of sheet erosion.

Rill Erosion

Rill erosion results when surface water runoff concentrates, forming small yet well-defined channels. These distinct channels where the soil has been washed away are called rills when they are small enough to not interfere with field machinery operations. In many cases, rills are filled in each year as part of tillage operations.

The distinct path where the soil has been washed away by
surface water runoff is an indicator of rill erosion.

Gully Erosion

Gully erosion is an advanced stage of rill erosion where surface channels are eroded to the point
where they become a nuisance factor in normal tillage operations. There are farms in Ontario that
are losing large quantities of topsoil and subsoil each year due to gully erosion. Surface water runoff, causing gully formation or the enlarging of existing gullies, is usually the result of improper
outlet design for local surface and subsurface drainage systems. The soil instability of gully banks,
usually associated with seepage of groundwater, leads to sloughing and slumping (caving-in) of
bank slopes. Such failures usually occur during spring months when the soil water conditions are
most conducive to the problem.

Gully formations are difficult to control if corrective measures are not designed and properly constructed. Control measures must consider the cause of the increased flow of water across the landscape and be capable of directing the runoff to a proper outlet. Gully erosion results in significant
amounts of land being taken out of production and creates hazardous conditions for the operators
of farm machinery.

Gully erosion may develop in locations where rill erosion has not been managed.

Bank Erosion

Natural streams and constructed drainage channels act as outlets for surface water runoff and sub-surface drainage systems. Bank erosion is the progressive undercutting, scouring and slumping of these drainage ways. Poor construction practices, inadequate maintenance, uncontrolled livestock access and cropping too close can all lead to bank erosion problems.

Bank erosion involves the undercutting and scouring of natural
stream and drainage channel banks.

Poorly constructed tile outlets also contribute to bank erosion. Some do not function properly because they have no rigid outlet pipe, have an inadequate splash pad or no splash pad at all, or have outlet pipes that have been damaged by erosion, machinery or bank cave-ins.

The direct damages from bank erosion include loss of productive farmland, undermining of structures such as bridges, increased need to clean out and maintain drainage channels and washing out of lanes, roads and fence rows.

Effects of Water Erosion

On-site

The implications of soil erosion by water extend beyond the removal of valuable topsoil. Crop emergence, growth and yield are directly affected by the loss of natural nutrients and applied fertilizers. Seeds and plants can be disturbed or completely removed by the erosion. Organic matter from the soil, residues and any applied manure, is relatively lightweight and can be readily transported off the field, particularly during spring thaw conditions. Pesticides may also be carried off the site with the eroded soil.

Soil quality, structure, stability and texture can be affected by the loss of soil. The breakdown of aggregates and the removal of smaller particles or entire layers of soil or organic matter can weaken the structure and even change the texture. Textural changes can in turn affect the water-holding capacity of the soil, making it more susceptible to extreme conditions such as drought.

Off-site

The off-site impacts of soil erosion by water are not always as apparent as the on-site effects. Eroded soil, deposited down slope, inhibits or delays the emergence of seeds, buries small seedlings and necessitates replanting in the affected areas. Also, sediment can accumulate on down-slope properties and contribute to road damage.

Sediment that reaches streams or watercourses can accelerate bank erosion, obstruct stream and drainage channels, fill in reservoirs, damage fish habitat and degrade downstream water quality. Pesticides and fertilizers, frequently transported along with the eroding soil, contaminate or pollute downstream water sources, wetlands and lakes. Because of the potential seriousness of some of the off-site impacts, the control of "non-point" pollution from agricultural land is an important consideration.

Wind Erosion

Wind erosion occurs in susceptible areas of Ontario but represents a small percentage of land – mainly sandy and organic or muck soils. Under the right conditions it can cause major losses of soil and property.

Wind erosion can be severe on long, unsheltered, smooth soil surfaces.

Soil particles move in three ways, depending on soil particle size and wind strength – suspension, saltation and surface creep.

The rate and magnitude of soil erosion by wind is controlled by the following factors:

Soil Erodibility

Very fine soil particles are carried high into the air by the wind and transported great distances (suspension). Fine-to-medium size soil particles are lifted a short distance into the air and drop back to the soil surface, damaging crops and dislodging more soil (saltation). Larger-sized soil particles that are too large to be lifted off the ground are dislodged by the wind and roll along the soil surface (surface creep). The abrasion that results from windblown particles breaks down stable surface aggregates and further increases the soil erodibility.

Soil Surface Roughness

Soil surfaces that are not rough offer little resistance to the wind. However, ridges left from tillage can dry out more quickly in a wind event, resulting in more loose, dry soil available to blow. Over time, soil surfaces become filled in, and the roughness is broken down by abrasion. This results in a smoother surface susceptible to the wind. Excess tillage can contribute to soil structure breakdown and increased erosion.

Climate

The speed and duration of the wind have a direct relationship to the extent of soil erosion. Soil moisture levels are very low at the surface of excessively drained soils or during periods of drought, thus releasing the particles for transport by wind. This effect also occurs in freeze-drying of the soil surface during winter months. Accumulation of soil on the leeward side of barriers such as fence rows, trees or buildings, or snow cover that has a brown colour during winter are indicators of wind erosion.

Unsheltered Distance

A lack of windbreaks (trees, shrubs, crop residue, etc.) allows the wind to put soil particles into motion for greater distances, thus increasing abrasion and soil erosion. Knolls and hilltops are usually exposed and suffer the most.

Vegetative Cover

The lack of permanent vegetative cover in certain locations results in extensive wind erosion. Loose, dry, bare soil is the most susceptible; however, crops that produce low levels of residue (e.g., soybeans and many vegetable crops) may not provide enough resistance. In severe cases, even crops that produce a lot of residue may not protect the soil.

The most effective protective vegetative cover consists of a cover crop with an adequate network of living windbreaks in combination with good tillage, residue management and crop selection.

Effects of Wind Erosion

Wind erosion damages crops through sandblasting of young seedlings or transplants, burial of plants or seed, and exposure of seed. Crops are ruined, resulting in costly delays and making re-seeding necessary. Plants damaged by sandblasting are vulnerable to the entry of disease with a resulting decrease in yield, loss of quality and market value. Also, wind erosion can create adverse operating conditions, preventing timely field activities.

Soil drifting is a fertility-depleting process that can lead to poor crop growth and yield reductions in areas of fields where wind erosion is a recurring problem. Continual drifting of an area gradually causes a textural change in the soil. Loss of fine sand, silt, clay and organic particles from sandy soils serves to lower the moisture-holding capacity of the soil. This increases the erodibility of the soil and compounds the problem.

The removal of wind-blown soils from fence rows, constructed drainage channels and roads, and from around buildings is a costly process. Also, soil nutrients and surface-applied chemicals can be carried along with the soil particles, contributing to off-site impacts. In addition, blowing dust can affect human health and create public safety hazards.

Tillage Erosion

Tillage erosion is the redistribution of soil through the action of tillage and gravity. It results in the progressive down-slope movement of soil, causing severe soil loss on upper-slope positions and accumulation in lower-slope positions. This form of erosion is a major delivery mechanism for water erosion. Tillage action moves soil to convergent areas of a field where surface water runoff concentrates. Also, exposed subsoil is highly erodible to the forces of water and wind. Tillage erosion has the greatest potential for the "on-site" movement of soil and in many cases can cause more erosion than water or wind.

Tillage erosion involves the progressive down-slope movement of soil.

The rate and magnitude of soil erosion by tillage is controlled by the following factors:

Type of Tillage Equipment

Tillage equipment that lifts and carries will tend to move more soil. As an example, a chisel plow leaves far more crop residue on the soil surface than the conventional moldboard plow but it can move as much soil as the moldboard plow and move it to a greater distance. Using implements that do not move very much soil will help minimize the effects of tillage erosion.

Direction

Tillage implements like a plow or disc throw soil either up or down slope, depending on the direction of tillage. Typically, more soil is moved while tilling in the down-slope direction than while tilling in the up-slope direction.

Speed and Depth

The speed and depth of tillage operations will influence the amount of soil moved. Deep tillage disturbs more soil, while increased speed moves soil further.

Number of Passes

Reducing the number of passes of tillage equipment reduces the movement of soil. It also leaves more crop residue on the soil surface and reduces pulverization of the soil aggregates, both of which can help resist water and wind erosion.

Effects of Tillage Erosion

Tillage erosion impacts crop development and yield. Crop growth on shoulder slopes and knolls is slow and stunted due to poor soil structure and loss of organic matter and is more susceptible to stress under adverse conditions. Changes in soil structure and texture can increase the erodibility of the soil and expose the soil to further erosion by the forces of water and wind.

In extreme cases, tillage erosion includes the movement of subsurface soil. Subsoil that has been moved from upper-slope positions to lower-slope positions can bury the productive topsoil in the lower-slope areas, further impacting crop development and yield. Research related to tillage-eroded fields has shown soil loss of as much as 2 m of depth on upper-slope positions and yield declines of up to 40% in corn. Remediation for extreme cases involves the relocation of displaced soils to the upper-slope positions.

Conservation Measures

The adoption of various soil conservation measures reduces soil erosion by water, wind and tillage. Tillage and cropping practices, as well as land management practices, directly affect the overall soil erosion problem and solutions on a farm. When crop rotations or changing tillage practices are not enough to control erosion on a field, a combination of approaches or more extreme measures might be necessary. For example, contour plowing, strip-cropping or terracing may be considered. In more serious cases where concentrated runoff occurs, it is necessary to include structural controls as part of the overall solution – grassed waterways, drop pipe and grade control structures, rock chutes, and water and sediment control basins.

SOIL CONSERVATION

Soil Conservation is the name given to a handful of techniques aimed at preserving the soil. Soil loss and loss of soil fertility can be traced back to a number of causes including over-use, erosion, salinization and chemical contamination. Unsustainable subsistence farming and the slash and burn clearing methods used in some less developed regions, can often cause deforestation, loss of soil nutrients, erosion on a massive scale and sometimes even complete desertification.

Soil erosion removes the top soil that is necessary for organic matter, nutrients, micro-organisms that are requires for plants to grow and shine. Soil conservation is one such step that protects the soil from being washed away. The soil then ends up in aquatic resources bringing in pesticides and fertilizers used on agricultural land. Healthy soil is important for plants to grow and flourish. Taking necessary steps to conserve the soil is part of environmentally friendly lifestyle. There are several ways to conserve soil that can be done through agricultural practices or measures you take at home.

Following methods are normally adopted for conserving soil:

- Afforestation: The best way to conserve soil is to increase area under forests. Indiscriminate felling of trees should be stopped and efforts should be made to plant trees in new areas. A minimum area of forest land for the whole country that is considered healthy for soil and water conservation is between 20 to 25 per cent but it was raised to 33 per cent in the second five year plan; the proportion being 20 per cent for the plains and 60 per cent for hilly and mountainous regions.

- Checking Overgrazing: Overgrazing of forests and grass lands by animals, especially by goats and sheep, should be properly checked. Separate grazing grounds should be ear-marked and fodder crops should be grown in larger quantities. Animals freely move about in the fields for grazing and spoil the soil by their hoofs which leads to soil erosion. This should be avoided.

- Constructing Dams: Much of the soil erosion by river floods can be avoided by constructing dams across the rivers. This checks the speed of water and saves soil from erosion.

- Changing Agricultural Practices: We can save lot of our valuable soil by bringing about certain changes in our agricultural practices. Some of the outstanding changes suggested in this context are as under.

- Crop Rotation: In many parts of India, a particular crop is sown in the same field year after year. This practice takes away certain elements from the soil, making it infertile and exhausted rendering it unsuitable for that crop. Rotation of crops is the system in which a different crop is cultivated on a piece of land each year.

This helps to conserve soil fertility as different crops make different demands on the soil. For example, potatoes require much potash but wheat requires nitrate. Thus it is best to alternate crops in the field. Legumes such as peas, beans, clover, vetch and many other plants, add nitrates to the soil by converting free nitrogen in the air into nitrogenous nodules on their roots.

Thus if they are included in the crop rotation nitrogenous fertilisers can be dispensed with. By rotating different types of crops in successive years, soil fertility can be naturally maintained. For example, wheat may be cultivated in the first year, barley in the second and legumes in the third.

The cycle may then be repeated. Further, there are some crops such as maize, cotton, tobacco and potato which can be classed as erosion inducing, whilst some other crops such as grass, forage crops and many legumes are erosion resisting. Small grain crops like wheat, barley, oats and rice are between these two extremes.

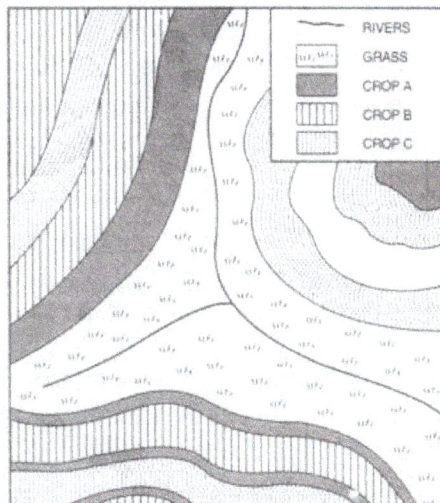

Strip farming following the contour pattern.

- Strip Cropping: Crops may be cultivated in alternate strips, parallel to one another. Some strips may be allowed to lie fallow while in others different crops may be sown e.g., grains, legumes, small tree crops, grass etc. Various crops ripen at different times of the year and are harvested at intervals. This ensures that at no time of the year the entire area is left bare or exposed. The tall growing crops act as wind breaks and the strips which are often parallel to the contours help in increasing water absorption by the soil by slowing down run off.

- Use of Early Maturing Varieties: Early maturing varieties of crops take less time to mature and thus put lesser pressure on the soil. In this way it can help in reducing the soil erosion.

- Contour Ploughing: If ploughing in done at right angles to the hill slope, following the natural contours of the hill, the ridges and furrows break the flow of water down the hill This prevents excessive soil loss as gullies are less likely to develop and also reduce run-off so that plants receive more water. Thus by growing crops in contour pattern, plants can absorb much of the rain water and erosion is minimised. When viewed from above, the field looks like a contour map.

Contour ploughing.

- Terracing and Contour Bunding: Terracing and contour bunding across the hill slopes is a very effective and one of the oldest methods of soil conservation. Hill slope is cut into a number of terraces having horizontal top and steep slopes on the back and front. Contour bunding involves the construction of banks along FIG Contour Ploughing the contours.

  Terracing and contour bunding which divides the hill slope into numerous small slopes, checks the flow of water, promotes absorption of water by soil and saves soil from erosion. Retaining walls of terraces control the flow of water and help in reducing soil erosion. Sometimes tree crops such as rubber are also planted to combat soil erosion.

  But there is a limit to which bunding is an effective measure of soil conservation. When the slope is steeper than 8 per cent or 1 in 12, bunding becomes expensive and less effective. Nothing over 20 per cent or 1 in 5 should be terraced. Fields of a slope steeper than 15 per cent or 1 in 6 should be withdrawn from ploughing as they are not usually worth the labour of making benches very close together.

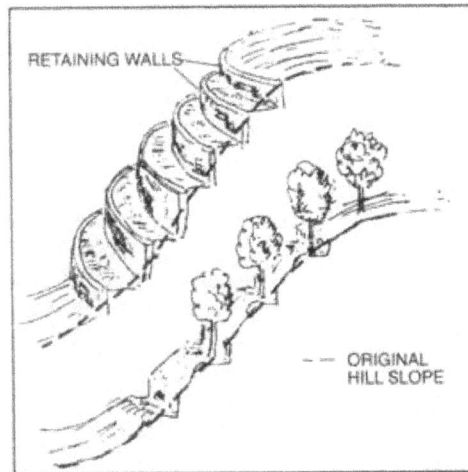

Two ways of terracing steep hill slopes to prevent soil erosion and
to produce flat-land conditions artificially.

- **Checking Shifting Cultivation:** Checking and reducing shifting cultivation by persuading the tribal people to switch over to settled agriculture is a very effective method of soil conservation. This can be done by making arrangements for their resettlement which involves the provision of residential accommodation, agricultural implements, seeds, manures, cattle and reclaimed land.

- **Ploughing the Land in Right Direction:** Ploughing the land in a direction perpendicular to wind direction also reduces wind velocity and protects the top soil from erosion.

References

- Soil-pollution, environmental-chemistry, chemistry, guides: toppr.com, Retrieved 21 August, 2019

- Sims, G.K. 2014. "Soil degradation", in accessscience@mcgraw-Hill, http://www.accessscience.com, DOI 10.1036/1097-8542.757375

- Causes-and-effects-of-soil-erosion, environment: eartheclipse.com, Retrieved 22 January, 2019

- Facts, engineer, English: omafra.gov.on.ca, Retrieved 23 February, 2019

- Methods-of-soil-conservation: conserve-energy-future.com, Retrieved 24 March, 2019

- Soil-conservation-4-methods-that-must-be-adopted-for-conserving-soil, geography: yourarticlelibrary.com, Retrieved 25 April, 2019

PERMISSIONS

All chapters in this book are published with permission under the Creative Commons Attribution Share Alike License or equivalent. Every chapter published in this book has been scrutinized by our experts. Their significance has been extensively debated. The topics covered herein carry significant information for a comprehensive understanding. They may even be implemented as practical applications or may be referred to as a beginning point for further studies.

We would like to thank the editorial team for lending their expertise to make the book truly unique. They have played a crucial role in the development of this book. Without their invaluable contributions this book wouldn't have been possible. They have made vital efforts to compile up to date information on the varied aspects of this subject to make this book a valuable addition to the collection of many professionals and students.

This book was conceptualized with the vision of imparting up-to-date and integrated information in this field. To ensure the same, a matchless editorial board was set up. Every individual on the board went through rigorous rounds of assessment to prove their worth. After which they invested a large part of their time researching and compiling the most relevant data for our readers.

The editorial board has been involved in producing this book since its inception. They have spent rigorous hours researching and exploring the diverse topics which have resulted in the successful publishing of this book. They have passed on their knowledge of decades through this book. To expedite this challenging task, the publisher supported the team at every step. A small team of assistant editors was also appointed to further simplify the editing procedure and attain best results for the readers.

Apart from the editorial board, the designing team has also invested a significant amount of their time in understanding the subject and creating the most relevant covers. They scrutinized every image to scout for the most suitable representation of the subject and create an appropriate cover for the book.

The publishing team has been an ardent support to the editorial, designing and production team. Their endless efforts to recruit the best for this project, has resulted in the accomplishment of this book. They are a veteran in the field of academics and their pool of knowledge is as vast as their experience in printing. Their expertise and guidance has proved useful at every step. Their uncompromising quality standards have made this book an exceptional effort. Their encouragement from time to time has been an inspiration for everyone.

The publisher and the editorial board hope that this book will prove to be a valuable piece of knowledge for students, practitioners and scholars across the globe.

INDEX

www.ingramcontent.com/pod-product-compliance
Lightning Source LLC
Chambersburg PA
CBHW082046190326
41458CB00010B/3471